'Frank, informative and often bleakly funny' Helen Brown, *Telegraph*

'Will interest anyone who wants to know what makes people tick . . . *The Incurable Romantic* earns its place in the fine tradition of popular psychoanalytic writing . . . an amiable and acute guide to the ma

'It is utterly compelling: the details, the dialogue, which bring each character, however heavily disguised, leaping off the page. Tallis's years of close observation might not always have solved his patients' problems ... but they have helped turn him into a fine writer ... He knows how to tell a story. Boy, does he know how to tell a story. This powerful and moving book is not just about individual cases. It's also about what the human animal needs ... They are certainly enough here to create something that feels profoundly truthful. Something that feels, in fact, like an act of love' Christina Patterson, *Sunday Times*

'[Tallis is] a brilliant raconteur with an acute ear for dialogue and sleuth-like capabilities. Only someone who has never felt sick falling head over heels, suffered the agonising pangs of jealousy, battled bestial fogs of lust or wallowed in the delirious happiness of being entwined with the object of their love could fail to be fascinated' *Evening Standard*

'A gifted storyteller ... Tallis's characters remain sharply, painfully real, their stories as inconclusive, messy and fascinating as life' *The Economist*

'Tallis's book reminded me of *Do No Harm* by the neurosurgeon Henry Marsh ... Through these cases Tallis makes a strong case that "love" can be the cause of great distress in many ways. He intersperses the cases with observations from history, literature and scientific reports, making for an enjoyable, entertaining and informative read' Richard Smith, British Medical Journal Opinion

'Tallis's book is about what happens when these perfectly ordinary feelings become warped, excessive and unmanageable in some of his patients, and as you can imagine, it's pretty gripping ... The great thing about *The Incurable Romantic* is that it makes you feel better about yourself. Whether you're happily or unhappily married, happily or unhappily single, involved in an adulterous relationship with another person or even several other people, you're doing better than these guys' Nick Hornby

THE INCURABLE ROMANTIC

and other unsettling revelations

FRANK TALLIS

ABACUS

8 by Little, Brown
1 2019 by Abacus

4 2

A CIP catalogue record for this book
is available from the British Library.

ISBN 978-0-349-14295-1

Typeset in Dante by M Rules
Printed and bound in Great Britain by
Clays Ltd, Elcograf S.p.A.

Papers used by Abacus are from well-managed forests
and other responsible sources.

Abacus
An imprint of
Little, Brown Book Group
Carmelite House
50 Victoria Embankment
London EC4Y 0DZ

An Hachette UK Company
www.hachette.co.uk

www.littlebrown.co.uk

To Nicola, incurably

Contents

Preface

The Roman philosopher Lucretius is famous for writing an extended poem titled *On the Nature of Things*. It contains sections on a wide range of subjects, such as the movement of atoms, the cosmos, time – and a great deal of psychology.

Among Lucretius's writings on the mind and behaviour is a description of what happens when people fall in love. He observes that the besotted frequently become agitated and stirred up by insatiable desires. Sexual union, often passionate and violent, results in only temporary relief, because lovers always want more of each other. Lucretius seems to be describing an addiction. He uses language that suggests that falling in love is a little like becoming ill or, even worse, going mad. Love, he says, is like an unconquerable disease and lovers waste away from wounds that can't be seen. They are lovesick: weak and neglectful of responsibilities, they behave foolishly and fritter away fortunes on excessive gifts; they become jealous and insecure.

After describing all these symptoms, Lucretius employs a device that many stand-up comedians use. He subverts our expectations to make us laugh. He says: That's what love's like when things are good – just imagine what it's like when things get bad. All of a sudden, he is no longer a classical philosopher but a friend or drinking companion.

Lucretius proceeds to tell us what happens when love goes wrong. Lovers become delusional and lose the power to make objective judgements. They experience a kind of ongoing hallucination. Ordinariness, or even ugliness, is perceived as outstanding beauty. They can't keep away from their beloved, and everyone else in the world becomes insignificant. Lovers become abject and helpless, and what pleasures they enjoy – sensuality, mutual delight – serve only to limit them. The goddess of love, Lucretius warns, has sturdy fetters.

It's interesting, even remarkable, that a Roman philosopher, dead for over two thousand years, can supply us with a description of lovesickness that we all recognise. In this respect, it doesn't seem that human nature has changed very much since classical times. But Lucretius doesn't stop there. He refines his argument and makes a distinction between love going well and love going wrong – love that is normal and abnormal love. In a more general sense, the whole discipline of psychiatry is predicated on this division: the identification of abnormal individuals within the wider, 'normal' population.

In fact, the symptoms that Lucretius associates with love going well are only marginally less dramatic than the symptoms he associates with love going wrong. This suggests a continuum of increasing severity, rather than a real difference between normal and abnormal. I doubt Lucretius had particularly strong views on this issue, and the distinction he makes might appear in his poem only in order to make his joke work.

Lucretius described the lovelorn as fools. Indeed, the tone of his verse is quite contemptuous. He invites us to laugh along with him at their folly. It's an attitude that many may share. There's a certain amount of questionable pleasure to be had from watching other people making fools of themselves, but when we mock the lovelorn, we do so either as hypocrites or automatons. Who hasn't acted foolishly – or at the very least conspicuously out of character – when in love? Only those who renounce society or repress their emotions are immune.

We know almost nothing about Lucretius. Saint Jerome tells us that he committed suicide when he reached his middle years. It is thought that he had been driven mad by a love potion. Perhaps he should have taken lovesickness more seriously.

She was clever, successful and horribly depressed – an opera singer with a very considerable talent. As is often the case with depressed patients, she was also extremely irritable. She told me what sex felt like with her husband: 'I feel like a blow-up doll,' she said, forming an 'O' with her mouth and stiffening her limbs. Then, suddenly, she looked at me as if she'd only just noticed I was sitting there. Her eyes narrowed. 'Why do you do this?' she demanded. My answer was thoughtless and trite. 'It's my job . . .' I should have known better and didn't get the chance to elaborate. She was expecting something more insightful from a psychologist. 'All this misery and unhappiness,' she exploded. 'Day after day – listening to people's shit – listening to my shit! What kind of person does this for a living?' Then the fire went out of her eyes and I could see her sinking into a quagmire of self-loathing. She made a feeble, apologetic gesture. 'It's okay,' I said. And I gave her a better answer – although it was still incomplete and a little disingenuous.

Why *did* I become a psychotherapist?

The saccharine and safe answer is that I wanted to help people. And that would be true. But this is so obvious as to be completely uninformative. A little like asking a fireman why he chose to join the fire brigade only to be given the flat answer, 'To put out fires.'

For as long as I can remember, I have always been attracted to hinterlands, fringes, twilight places and oddity. As an adolescent I would consume volumes of weird fiction and horror, largely because these genres typically explored the darker recesses of the mind and bizarre behaviour. As I matured, this fascination with oddity (and particularly psychological oddity) became something less prurient and somewhat closer to intellectual curiosity. But it remained, in essence, unchanged.

I've worked in many different settings, including in some very large, rambling hospitals. In every instance, when the opportunity arose, I would escape the busy, pristine 'front of stage' areas – reception, outpatients, wards – stray off the major thoroughfares and wander around basements, neglected corridors and empty offices. Sometimes I would stroll through eerie, silent places for some time without encountering another soul. On one of my excursions, I found what appeared to be an abandoned operating theatre with a ceiling constructed of glass panels. Much of the glass was broken and autumn leaves were scattered on the tiled floor. In the centre of the space was an antiquated machine with white enamelled surfaces. It was vaguely telescopic, mounted on a wheel-shaped base and festooned with levers. I felt as if I had stepped into a novel by H. G. Wells or Jules Verne. On another occasion, I discovered a room lined with dusty shelves and on each of these were rectangular Perspex containers in which slices of human brain were preserved in formaldehyde. It was a haunting image – like a library of memories. In the grounds of a Victorian asylum I came across a tiny museum that contained a collection of art works by former patients. I was the only visitor. A custodian appeared – a diminutive, alert woman – who immediately demanded to hear my views on the effect of hot weather on homicidal behaviour.

Symptoms must have causes. They can be produced by abnormalities in the brain, neurotransmitter imbalances, repressed memories, or distorted thinking. But symptoms are also the end point of stories. For me, psychotherapy is as much about narrative as it is about science or compassion, perhaps even more so. The awkward truth, which I couldn't reveal to the depressed opera singer, was that I found the day-to-day misery of psychotherapy tolerable because I liked listening to the stories – especially those that were touched by strangeness and explained the occurrence of unusual or striking clinical presentations. My uneasy conscience is salved, in this respect, by the fact that I stand shoulder to shoulder with some very august company.

The practice of psychotherapy has long been associated with storytelling. Anna O., the very first patient to be treated using a procedure that eventually became psychoanalysis, entered an altered state of consciousness during which she would tell Josef Breuer (Freud's avuncular patron and collaborator) stories that reminded him of those written by Hans Christian Andersen. These formed an integral part of her treatment and prompted her to describe Breuer's approach as the 'talking cure'.

People are living story books. Talking cures open the covers and let the stories out.

The core of this book is a series of true stories about real people, all of whom I saw for psychotherapy because they experienced significant distress attributable to falling in love or being in love. Most of their problems were emotional, sexual, or a combination of the two. Romantic love, as Lucretius suggests, is almost always linked with physical desire. The clinical phenomena I describe (the symptoms, feelings and behaviours) are authentic; however, I have disguised my patients to ensure anonymity.

The very earliest poems were composed in Egypt over three and a half thousand years ago – exquisite love songs that describe the despair of lovers as a malady. Early medical texts also conceptualise falling in love as an ailment. The second-century Greek physician Galen described a married woman who couldn't sleep and who started acting strangely because she had fallen in love with a dancer. Lovesickness was considered a legitimate diagnosis from classical times to the eighteenth century, but it more or less disappeared in the nineteenth century. Today, the term 'lovesickness' is employed as a metaphor rather than a diagnosis.

When love-struck individuals voice their complaints, the best they can usually hope for is a little sympathy and a wry, knowing smile. Teasing and ridicule are also common responses.

But lovesickness is not a trivial matter. Unrequited love is a frequent cause of suicide (particularly among the young) and approximately 10 per cent of all murders are connected with sexual jealousy. Moreover, there is a view that intermittently gains currency within psychiatry and psychology that troubled close relationships are not merely associated with mental illness but are a primary cause.

I have often found myself sitting in front of lovesick patients whose psychological pain and behavioural disturbances were equal in severity to any of the cardinal symptoms of a major psychiatric illness. Such patients are usually embarrassed to disclose their thoughts and feelings, having internalised the prevailing view that lovesickness is transitory, adolescent, inconsequential, or ridiculous. This couldn't be further from the truth. The emotional and behavioural consequences of falling in love can be enduring and profound. I have seen conventional lives unravel on account of wild passions; I have watched people suffer prolonged agonies because of rejection; I have accompanied individuals to the verge of psychological precipices – dark, fearful places – where I sensed that an infelicitous word or maladroit turn of phrase might be enough to propel them over the edge; I have seen patients listening to the siren call of oblivion, attending to its promises of release and eternal rest, even as I endeavoured, sometimes desperately, to persuade them to step back. I have watched people hollowed out by desire and yearning diminish to a fading iteration of their former selves. On none of these occasions was I ever tempted to offer a wry, knowing smile.

The term 'incurable romantic' is more than just an amusing designation – it acknowledges an uncomfortable clinical reality. One of the ardent poets of ancient Egypt tellingly wrote that doctors with their remedies could not heal his heart. He may have been right.

Love is a great leveller. Everyone wants love, everyone falls in love, everyone loses love and everyone knows something of love's madness; and when love goes wrong, our relative wealth, education and status count for nothing. The jilted lord is just as vulnerable as the jilted bus driver. Virtually all the major theoreticians of psychotherapy, from Freud onwards, agree that love is essential to human happiness.

It is my belief that the problems arising from love – infatuation, jealousy, heartbreak, trauma, inappropriate attachment and addiction, to name but a few – merit serious consideration and that the line which separates normal from abnormal love is frequently blurred. I hope that this view will be supported by the sometimes quite unsettling revelations that follow – unsettling, because ultimately, they demonstrate the presence of deep-rooted and universal vulnerabilities that have been locked into our nervous systems by evolutionary processes. The merest spark of sexual attraction can cause a fire that has the potential to consume us. We all share this dormant propensity, which explains why examples of its full expression in the clinic are so arresting and alarming. They give us good reason to reflect on our own intimate histories and forewarn us of dangers that may lie ahead.

Psychotherapy is a notoriously divided discipline. There are many different schools of thought (e.g. psychoanalytic, gestalt, rational-emotive) and each of these schools is represented by figureheads whose particular approach – while preserving a circumscribed set of basic values and principles – diverges from the mainstream. These departures from orthodoxy range from minor modifications of theory to major doctrinal revisions. The history of psychotherapy is one of internecine strife, schisms, secession and intellectual hostility. One can imagine it represented on a page as a complex tree diagram composed of several trunks and each of these trunks producing numerous branches and offshoots. This process of growth and repeated bifurcation has taken place over a period of just over a hundred years and continues to this day.

It is customary for a book of this kind to reflect the theoretical orientation of the author. Typically, symptoms are interpreted and understood within the context of the author's favoured unitary approach. I have always found allegiance to a single school of psychotherapy unnecessarily limiting as I believe that even the most peripheral innovators in the history of the subject have had something significant or useful to say about the origin, maintenance and cure of symptoms. Thus, the clinical descriptions in this book are presented with commentaries that borrow from many different perspectives.

While psychotherapists have been engaged in their various disputes with each other, they have also been participating – as a more unified group – in a much bigger, ongoing dispute with biological psychiatrists concerning the ultimate origin of mental illness. Biological psychiatry is based on the assumption that all mental illnesses are caused by structural or chemical abnormalities in the brain. A corollary of this assumption is that biology, being a more fundamental science, trumps psychology. The relative status afforded to biological and psychological accounts of mental illness frequently polarises views, and opponents from both camps are usually committed and vociferous. Once again, I find this debate – in its extreme form – rather sterile.

Even if one supposes that all mental states can be mapped onto brain states, this doesn't mean psychology is invalidated, in the same way that biology isn't invalidated by chemistry, and likewise, chemistry isn't invalidated by physics. Almost everything in the universe can be described in different ways and at different levels, and the mental life of humans is no exception. Multiple perspectives are illuminating and give a more complete and satisfying account of phenomena. Consequently, my case commentaries also include references to biological psychiatry and brain sciences.

He was nineteen, a philosophy student with unwashed hair and an unconvincing beard. The dark crescents under his eyes suggested sleepless nights and his clothes exuded the smell of cigarettes. He had been rejected by his girlfriend and he was exhibiting many of the symptoms of lovesickness described by poets through the ages. His distress and anger seemed to come off his body in rising waves.

'I don't understand how it happened. I just don't understand. ' I noticed his foot tapping impatiently. 'Can you give me *any* answers?' His emphasis converted an innocent question into a challenge that also carried with it a subtle slur, the imputation of impotence.

'That rather depends on your questions,' I replied.

His pale cheeks acquired some colour. 'What's it all about? I mean . . . life, love. What's it all about?'

Love and life are often linked together because it is almost impossible to think about life without love. In a very real sense, when we ask questions about the nature of love we are also asking very deep questions about what it is to be human and how to live.

My young patient threw his arms out and kept them suspended in the air: 'Well?'

Chapter 1

The Barristers' Clerk:
Love that accepts no denial

We were seated on two high-backed armchairs, facing each other across a small table. Within easy reach was the indispensable tool of the professional psychotherapist, a box of tissues – perhaps the most underwhelming of all occupational accessories. I've spent many, many hours of my life watching people cry.

Megan was a woman in her mid-forties, conservatively dressed, with soft, rounded features. Her hair was dark brown and styled in a neat bob – the straight sides curled inwards under her chin. She had a kind face. In repose, her features retained the suggestion of a deferential, self-conscious smile. The hem of her skirt descended some way below her knees and her shoes were of the sensible variety. An uncharitable person might have described her as dowdy.

Her GP had sent me a referral letter summarising the key facts of her case. Referral letters (typically dictated onto a recording device and later transcribed by a secretary) are neutral in tone. The short, clipped sentences tend to stifle drama: name, age, address, circumstances. Yet, Megan's history had retained its theatrical heat. The GP's bullet-point delivery had failed to refrigerate the essential elements of a tragic love story: emotional extremity, reckless abandon, passion and desire.

Before Megan stepped into my consulting room I'd studied the referral letter and, naturally, wondered what she would look like. My brain was quick to cast a suitable romantic heroine. I had imagined someone lean, tall, with wild hair and haunted eyes. I have to admit I was a little disappointed when Megan walked in.

At some level all clichés are true and appearances *can* be very deceptive. We rarely see each other when we first meet. It takes a lot of looking to see who is really there. At that early juncture, I could see only a barristers' clerk. In reality, the creature sitting in front of me was far more exotic, but I couldn't see past the obstruction of my own prejudices.

After a few introductory remarks, I explained that I had read her doctor's referral letter. Nevertheless, I still wanted to hear her version of events.

'It's difficult,' she said.

'Yes,' I agreed. 'I'm sure it is.'

'I can tell you things,' she continued. 'I can tell you what happened – but it's so difficult to express how it feels.'

'There's no rush,' I replied. 'Just take your time.'

Other than a few episodes of mild depression, Megan had never suffered from any significant psychological problems. 'My depression was never very serious,' she said. 'I mean, not like some people I know. I just used to get a bit moody, that's all. And after a few weeks my mood would lift and I'd feel okay again.'

'Did you identify any triggers?'

'The barristers I work for can be demanding. Perhaps it was stress.'

I nodded sympathetically and made a few notes.

Megan had been married for twenty years. Her husband, Philip, was an accountant and they had always been happy together. 'We don't have any children,' she volunteered. 'It isn't that we made a decision not to have children – it was just never the right time. We kept putting it off until eventually it didn't seem an issue any more. Sometimes I wonder what it would have been like, to have had children, to be a mum, but I can't say it's one of life's big regrets. I don't think I've missed out. And I'm sure Phil feels the same way.'

Two years earlier Megan had had to consult a dentist who specialised in complicated extractions.

'Can you remember meeting him for the first time?'

'Daman?' Her use of the dentist's given name was a little unusual. It needn't have been significant, but in this instance it was.

'Mr Verma.' I wasn't correcting her, merely confirming that we were talking about the same person.

She looked at me quizzically and I made a small gesture, encouraging her to continue. 'He examined me – told me that I should have the tooth removed – and I went home.'

'Did you find him attractive, did you *feel* anything?'

'I thought he was quite handsome. He had a pleasant manner. But . . .' She shook her head. 'I don't know. You see, this is why it's so difficult. These things are so hard to describe. Perhaps I felt something – right at the beginning. Yes. I probably did. I just wasn't sure what was happening. I was confused.'

I detected a note of distress in her voice. 'It's okay . . .'

Daman Verma performed the operation. There were no problems and everything went to plan. When the general anaesthetic wore off and Megan woke up, she felt different. 'I was aware of people moving around me – the two nurses . . . There were sounds, voices. I opened my eyes and looked up at a light on the ceiling and I remember thinking: I've got to see him. I wasn't frightened or worried. I didn't want to know how the operation had gone. All that I wanted was to see him.'

'Why?'

'I just . . . had this need. It felt – I don't know – necessary.'

'Did you want to say something to him?'

'No. I just wanted to see him.'

'Yes, but why?' I pressed her for a more precise answer but she was either unwilling or unable to give me one.

The dentist was called and he came to the recovery room. He held Megan's hand and probably said some words of reassurance. She couldn't remember, because she wasn't really listening. She had become entirely absorbed by his face, which struck her as being unnaturally beautiful, a face that in her view expressed the prime virtues of masculinity – strength, competence, accomplishment – and she discovered in his eyes something quite extraordinary, something that was so unexpected that it almost made her gasp: mutuality, reciprocation. He wanted her as much as she wanted him.

It was obvious. Why hadn't she seen it before? When he tried to move away she gripped his hand a little tighter. He looked embarrassed. Of course, he *would* be embarrassed. He couldn't show his feelings, not there, not in front of the nurses. How could he make a declaration of love in the recovery room? He had his reputation to consider, he was a professional. She was mildly amused by his play-acting, his clumsy attempts to conceal the truth. She released his fingers, knowing, with absolute certainty, that the love they felt for each other was so strong, so utterly overwhelming, that they would spend the rest of their lives together and very likely die together.

A princess wakes from a deep, enchanted sleep and gazes into the eyes of her Prince Charming. This scene appears in *Little Briar Rose* by the Brothers Grimm, but the Brothers Grimm were preceded over a hundred years earlier by Charles Perrault, who wrote *The Sleeping Beauty*.

Is it possible to fall so deeply in love, so quickly? Or is that something that happens only in fairy stories? Judgements concerning attractiveness are made in a matter of milliseconds, and, if they are positive, they are followed by congruent inferences. We assume that beautiful people are more likeable, friendly and interesting. It's a well-documented phenomenon that psychologists call the Halo Effect. Megan, however, had experienced something much more profound. It seems improbable that strangers can form an instant, meaningful and enduring bond. How can it possibly work out? The parties don't know each other. Yet, a high proportion of the general population claim to have experienced love at first sight, and many love-struck couples stay together. Some psychologists have suggested that instant attraction confers certain evolutionary advantages. For example, it hastens sexual contact so fewer opportunities for reproduction are wasted. This increases the probability of genes being transferred to the next generation, which is good for the individual (or at least their genes) and ultimately good for the species. Proneness to falling in love at first sight might be a very fundamental biological predisposition.

The fact that Megan started falling in love with Verma the instant she met him might not be so very remarkable; however, her insistence that her feelings were reciprocated was something quite different, as was her certainty. People often talk about being on the same wavelength and knowing each other's minds, but few would say that they have *certain* knowledge of someone else's thoughts and feelings, especially after such a short acquaintance.

'How did you know that Daman Verma had fallen in love with you?'

'I just knew.'

'Yes, but how?'

'I just knew.'

The repetition of this single phrase created a conversational stockade. I paused to consider how I might best negotiate the impasse. From Freud's time to the present, psychotherapists have made much use of a technique known as Socratic questioning. It is used to challenge assumptions and help patients think more critically. Socratic questioning tends to work best when it isn't interrogative, but gentle and oblique. The approach is consistent with a nugget of oriental wisdom that advises: 'Flow around obstacles, don't confront them.'

'Why is it,' I asked, 'that we believe some things and not others?'

Megan squinted at me as if I'd suddenly gone out of focus. 'Because we have reasons . . . '

'So what were your reasons; your reasons for believing that Daman Verma had fallen in love with you?'

'It's not something you can analyse.'

'Perhaps you're right. But I'd still like to talk about this for a while. Just to see if we can learn anything?'

Megan remained silent. Sometimes – during therapy – a silence descends that seems to arrest the passage of time. Everything becomes still. So still, in fact, that even asking a question seems clumsy and coercive. I changed position. This simple expedient broke the spell and time began to flow again.

'I could see it in his eyes.'

'What could you see?'

'His need. You can see things in people's eyes, can't you?' Defensiveness had made her voice brittle.

'We interpret expressions all the time. But do we really know what someone's thinking just from how they look?'

'Not always.'

'You were Daman Verma's patient and you'd asked to see him. Is it possible that you misinterpreted his expression? That what you saw was actually something closer to caring or concern?'

'What I saw was more meaningful. They say there's a look – you know? – the look of love . . .'

People do indeed talk about the look of love. What they are actually referring to is something that scientists call the copulatory gaze: the eyes lock for several seconds before one party looks away. It occurs when prospective lovers first encounter each other and this intense, probing stare usually signals sexual interest. Apes do much the same thing.

'You're certain.'

'Yes.'

'There aren't any alternative explanations?'

'No, not really . . .'

'It was in his eyes.'

'I know what I saw.' She raised her hands, showed me her palms and gave me an apologetic smile. What was she supposed to say?

In reality, there had been nothing exceptional in Verma's eyes. Not even the faintest glimmer of desire. Megan was just another patient. He was a busy dentist with several affiliations and a large private practice. As far as he was concerned, they had met, he had operated on her and now they would part company. When he left the recovery room, he might have reasonably supposed that, apart from a few follow-up appointments, he would never see her again. But if he did think that, his supposition would, in the fullness of time, be proved wrong. Very wrong.

'I couldn't stop thinking about him. And I could sense him thinking about me.'

'What do you mean? *Sense* . . .'

Megan ignored my question. 'It was so unfair. We both wanted to be together, but he couldn't work out how to deal with his situation.'

'If he'd really wanted to be with you, wouldn't he have left his wife?'

'No. He's a kind person – a really kind person. He didn't want to hurt her feelings.'

'Did he ever say that to you?'

'He didn't need to.' She looked at me with a weary expression. It was obvious that she didn't want to justify herself again. Even Socratic questioning gets tiresome.

After her operation, Megan obsessed about Verma day and night. Her sleep was disturbed and when she returned to work she couldn't concentrate. She yearned to be near him.

'Was the attraction sexual?'

'No,' she protested. Then she sighed. 'Well, yes. That was part of it. But it was only a small part. It's misleading – sex. I mean, if it had been possible for us to be together, and the physical side hadn't happened, that wouldn't have mattered. Not really. We'd have still wanted each other.'

Her husband had noticed that her mood was deteriorating. There was no obvious cause. He tried talking to her, but she was distant and withdrawn.

Weeks passed.

Megan's desire to contact Verma mounted day by day. Separation was becoming intolerable, a kind of torment. She found the courage to telephone him. 'It was an awkward conversation. I gave him a chance to tell me how he felt but he was obviously scared. The experience had been too overwhelming for him.'

'What did you talk about?'

'At first we talked about my recovery – how it was going. Eventually I had to say something more direct. I suggested that we meet up for a coffee, to discuss what we were going to do. Temple isn't that far from Harley Street. I said I'd get a cab.'

'And how did he respond?'

'He pretended he didn't understand. I persevered, but he was evasive. He made some excuse and hung up.'

'He was frightened by his own feelings and had to end the call.'

'Exactly . . .'

'Is that the only interpretation?'

She shrugged.

Megan wasn't discouraged. She phoned Verma repeatedly, sometimes several times a day. The dental secretaries became frosty and asked her to stop. After conducting a little detective work she was able to obtain his home number. When his wife, Angee, picked up the phone, Megan did her best to explain the situation as sympathetically as she could – because that's what Daman would have wanted – but the dentist's wife became irritable.

'She told me to get help.'

'What did you think of that?'

'I was expecting it.'

'So you could see how your behaviour might have looked to others?'

'Mad, you mean?'

'I didn't say that.' I was being disingenuous. That's exactly what I meant.

'Yes,' she nodded. 'I could see . . .'

'Didn't that make you pause to reflect – reconsider what you were doing?'

'It wasn't important to me what other people thought.'

'What about now? Does it matter now?'

We stared at each other across the small table.

Megan wrote letters to Verma every day; long, detailed letters suggesting solutions, begging him to recognise that their love could not be disowned or denied. He would never be happy unless he accepted the truth. What was the point of pretending otherwise? He wasn't to blame, neither of them was to blame, how could they be? Something remarkable had happened, something wonderful and miraculous, and there was no going back. They had to be brave and embrace their future together. Their lives would never be the same. And if they attempted to live apart they would live as shadows, wretched and incomplete. And it wasn't only their future at stake. They had to think about their spouses' futures too. It was wrong to deceive Philip and Angee, to perpetuate a lie. They were good people and deserved more than a sham marriage.

'I waited outside his practice. I waited for hours. And when he came out I ran over to him.'

She paused and bit her lower lip.

'What happened?'

'He didn't want to talk. I told him I understood, that it was all happening so fast that maybe he needed more time. But in the end I said to him you're going to have to accept that this is real.'

Verma contacted Megan's GP, who contacted Megan's husband later the same day.

'What did Philip say when he found out what you were doing?'

Megan looked at the ceiling and placed her fingers over her mouth. Her speech was muffled but still intelligible: 'He wasn't very happy.'

What was wrong with Megan? Before meeting Daman Verma, her life had been fairly routine – a steady job, holidays and hobbies, the company of her husband. All that had suddenly changed.

Megan was suffering from a rare but well-documented mental illness called de Clérambault's syndrome, which was first described in detail by the French psychiatrist Gaëten de Clérambault in 1921. Typically, the affected individual, usually a woman, falls in love with a man (with whom she has had little or no prior contact) and comes to believe that he is also passionately in love with her. In many instances, the sufferer alleges that it was the man who fell in love first. This perception arises in the absence of any actual stimulus or encouragement. The man – sometimes also referred to as the victim or object – is often older, of higher social status, or a celebrity. His inaccessibility may act as a spur. A hapless and unwelcome pursuit follows which is experienced by the victim as extreme harassment. Men can also develop de Clérambault's syndrome, although women are much more vulnerable. The exact ratio isn't known, but it is thought to be about three to one.

De Clérambault's syndrome (or something very much like it) has been described for centuries and one can find similar cases in works dating back to classical times. So, when he wrote about it in 1921, de Clérambault was not breaking new ground as such, but merely revisiting a condition that had previously been called erotomania. Still, it was his name that became most strongly associated with what is undoubtedly the sovereign affliction among the maladies of love – particularly so in the latter half of the twentieth century. Perhaps this is because his description was more comprehensive, insofar as he emphasised emotional as well as sexual aspects of the condition. In the eighteenth century, for example, erotomaniacs were defined as 'Those who engage in the furious pursuit of vagrant or illicit lust'.

Today, the terms de Clérambault's syndrome and erotomania are used interchangeably. At one point, the condition attracted the somewhat insensitive appellation 'old maid's insanity'. In modern diagnostic systems, it has become Delusional Disorder: Erotomanic Type. Even so, de Clérambault continues to haunt the marginalia of psychiatry and many continue to use 'de Clérambault's syndrome' instead of the more correct contemporary alternative, probably because it sounds more pleasing and carries a suggestion of drama. It recalls an exciting period in the past when the mind was a dark continent and largely unexplored.

De Clérambault's most famous case was a 53-year-old French dressmaker who believed that King George V was in love with her. She visited England several times in order to pursue him and waited outside Buckingham Palace. When she saw a curtain move she concluded that the King was sending her signals. The fact that the King wasn't very forthcoming didn't alter the dressmaker's belief. She concluded that he was in a state of denial: 'The King might hate me, but he can never forget. I could never be indifferent to him, nor he to me.'

The dressmaker also suffered from a secondary illness, paranoid psychosis. She believed, for example, that the King sometimes meddled in her affairs. De Clérambault's syndrome is frequently associated with conditions such as schizophrenia or bipolar disorder. What made Megan so interesting was her ordinariness. There was nothing about her life, character or history which offered the slightest indication of what was to follow. She was proof that, as far as mental health is concerned, we all walk a tight-rope and it really doesn't take very much to make us lose balance and fall.

In addition to being awarded medals for distinguished service in the First World War, de Clérambault was also fêted as a significant artist. Some of his paintings are exhibited in French museums. His most original work is a series of photographic studies of women dressed in veils. While assigned to a military hospital in North Africa he discovered traditional Moroccan garments and became fascinated by drapery as an artistic subject. A traditional Freudian would find the symbolic implications of such an interest telling: concealment, temptation, unwrapping and the promise of revelation. They are strange, uncanny images, vaguely reminiscent of Victorian spirit photography and largely overlooked by cultural historians until only recently.

In 1934, after two unsuccessful cataract operations, de Clérambault sat in front of a mirror and shot himself with his old service revolver. His camera was focused on his own reflection.

He had composed a suicide note in which he endeavoured to explain his behaviour. It had been suggested that a painting he wished to bequeath to the Louvre had been fraudulently acquired in a sale. He had been dishonoured and an episode of melancholia followed. In reality, the prospect of going blind was probably the most significant factor. For years he had studied people from two simultaneous perspectives – with the eyes of an artist and a psychiatrist. He would have registered every swathe, fold and wrinkle of the social fabric and been able to determine what lay beneath. Life without such acute powers of perception wasn't worth living. He must have been looking closely at himself when he pulled the trigger. I wonder what he saw.

'How did Philip react?'

'He was upset. But he didn't say nasty things – he didn't accuse me of betraying him. We talked and I tried to explain, but he didn't understand. Not truly. He told me he loved me – and said he'd always be there for me. It was sad.'

'Because you didn't love him any more . . .'

Megan looked at me aghast: 'No, no. I've always loved Phil. It's just what I feel for Daman . . .' Her sentence trailed off and she looked around the room as if she'd lost something. Then her features hardened around a direct, unnerving stare. 'It's something else – something higher.'

'More spiritual?'

'I don't know, maybe. I'm not sure where I stand where God's concerned. But I do know it feels different to loving Phil, stronger, deeper – like something that was meant to be.'

'Fated?'

'Yes. That's the word. Fated . . .'

Megan was taken by her husband to see a psychiatrist who decided to put her on Pimozide, an anti-psychotic drug that reduces delusional thinking. It works by blocking dopamine receptors in the brain. The action of the neurotransmitter dopamine has been associated with numerous aspects of behaviour, everything from remembering to vomiting, but there is also a large body of evidence showing that it mediates pleasure and pleasure-seeking. Not surprisingly, it is thought to have an important role in the development of addictions. The dopaminergic circuitry of the brain has also been implicated in biological accounts of what we call romantic love.

Megan took her medication as instructed, even though she wasn't convinced that her love for Verma was, as the psychiatrist had suggested, the symptom of an illness. The drug had no effect. She felt just the same. The dose was subsequently increased – and still there was no effect. In fact, Megan's longing seemed to be getting more intense. She waited outside the dentist's practice with increasing frequency. Sometimes he would see her and send his secretary out with a message: *go home*. Megan didn't argue. What was the point? She smiled, nodded and made her way back to the tube station. It didn't matter, not in the grand scheme of things, because ultimately, her patience would be rewarded. On many occasions she escaped Verma's notice by hiding in a doorway or standing behind a parked van. Then her vigils might last all day. During the winter months, even when the temperature plummeted, she was warmed by the simple fact of Verma's proximity.

One late afternoon – around five o' clock – she observed him leaving his practice and followed him home. She stood beneath a lamp-post, opposite his front door, picturing him inside. When she was discovered by his wife, Angee, who just happened to look out of an upstairs window, Verma stormed out of his house and confronted Megan. He was angry and threatened to call the police. Megan found his performance inauthentic: 'He was pretending, for his wife's sake. Really, in his heart, he *wanted* me to be there.' Megan didn't put up any resistance. Whenever she was ordered to go home, she did so, but by this time her behaviour was making everyone – particularly Angee – nervous. The Vermas had two children, a boy aged eight and a girl aged ten, and Angee was worried about their safety. To his enormous credit, Daman Verma never called the police. He recognised that Megan was ill and acted accordingly. His wife, however, was less understanding.

'I know I caused him problems,' said Megan. 'And I'm really sorry about that. I wasn't trying to break up his marriage – because in a sense it was already over. I just wanted things to move on.'

In Ian McEwan's novel *Enduring Love*, the protagonist's relationship begins to fail when he is stalked by a de Clérambault sufferer. This is exactly what happened to Angee and Daman Verma. Neither of them could cope with the stress. They began to have arguments about what measures should be taken to stop Megan. In due course, Verma opted for a radical solution. He applied for a job in Dubai. The move wasn't entirely provoked by Megan. It was something that the Vermas had discussed before;

however, Megan's harassment certainly made the decision easier. Daman Verma had recognised that Megan's fierce, pathological love would never die. Ironically, what we call true love is nowhere near as durable as its pathological variant. Only by interposing a substantial distance between himself and Megan did Verma stand a chance of resuming a normal existence.

Daman Verma and his family had been living in Dubai for six months when Megan was referred to me. She was no longer under the care of a psychiatrist, and her GP believed that she was much improved. Nevertheless, he thought it would be helpful if she was given the opportunity to talk about her experiences with a psychotherapist. She had been traumatised, and like most trauma victims, she would make a better adjustment if she could make sense of her history. But the more I talked to Megan, the more I suspected that she wasn't very much improved at all. She'd simply become better at hiding her pain.

'You still miss Daman, don't you?'

'Yes. I miss him a lot.' Megan was studying her hands. Her head was bowed and she didn't make eye contact. 'I often think about what he's doing. You know – in Dubai . . . I think of him waking up and getting out of bed, brushing his teeth and going to work.' It was interesting that she didn't see him surrounded by his family. 'I imagine him in his car, driving, listening to the radio – the sun shining. I imagine him arriving at his new dental practice and getting ready for his patients. I see him – like I'm watching a film, or documentary – scrubbing, changing into his surgical gown.' Her finger tips touched. 'I like to be on my own in the early evening, because I know that in Dubai he's just gone to bed and he'll be lying in the dark without any distractions. And it's then I feel most that I can reach out to him, and he'll know that I'm thinking about him – and then he'll start thinking about me – and we'll both be thinking of each other – and it's like . . .' She raised her head and her expression was beatific – like a religious visionary. Her eyes were gleaming and her face was flushed. She was slightly breathless when she added: 'It's like we're one.'

I have no doubt that Megan's merging fantasies produced an ecstatic state similar to that described by mystics. The experience of the soul's reunification with God is heady and rapturous. So much so that erotic allegory is often employed in scriptures and religious poetry to capture the intensity of heavenly communion. Orgasm seems to provide the only serviceable precedent.

Freud borrowed the term 'oceanic feeling' from one of his correspondents to describe sensations of pleasurable dissolution; however, he never considered the phenomenon anything more than a psychological reversion to the primitive. Indeed, he believed that all symbiotic feelings are influenced by memories formed in early infancy, when the boundary that separates the ego from the rest of the world is still incomplete and porous. In a sense, the ecstasy of lovers and mystics refers back to the womb and breast-feeding. Perhaps we are always striving to recover something of our original state, which was blissfully free from the terrors of isolation. It is often said that we are born alone and die alone (an aphorism attributed to sources as diverse as the fourth-century BCE Indian philosopher Chanakya and the actor Orson Welles). That isn't strictly true. None of us are born alone – and perhaps we never forget it.

A delusion is a rigidly held belief that is maintained even when there is no evidence to support it; however, what constitutes good evidence differs from person to person. Megan regarded her own feelings as acceptable evidence. This had the effect of strengthening her beliefs. Daman Verma was in love with her. She knew that he loved her because she felt it so deeply – and strong feelings always mean something. The opposite is probably closer to the truth. Feelings are often vague, misleading and inconsistent. They don't always provide us with reliable information about the world, other people, or our circumstances.

I once treated a woman who was terrified of walking. There was nothing wrong with her legs or sense of balance: she was just scared of putting one foot in front of the other in order to get around. She had concluded that walking *was* dangerous because it *felt* dangerous.

It's frustrating when a patient doesn't get better. I was working on the assumption that if I kept on questioning Megan's rigid beliefs about Daman Verma, they might change. But that wasn't happening. My impatience made me more direct, less Socratic.

'Does it look like Daman loves you?'

'I think he does.'

'Still . . .'

'Yes.'

'He moved to Dubai. He's moved thousands of miles away.'

I let the words resonate in the silence that followed. And then I let the silence thicken and become coercive. Could she hear a whistling in her ears? The accelerating beat of her heart? Silences – long silences – can be very uncomfortable. They make demands. Megan looked at me, a little puzzled, almost certainly hurt.

Many years ago, I attended a psychoanalytic case meeting and the subject under discussion was how it is sometimes necessary for a therapist to let silences curdle. A colleague said: 'Therapy. It's like a pressure cooker: if you don't have pressure the food won't cook.' But it's difficult to watch a patient stewing.

Megan finally spoke. 'He doesn't want to upset his wife.' It had become something of a mantra.

The next time I saw her, Megan looked more tired than usual.

'I wish I could talk to him on the phone,' she admitted. 'Even if it was for five minutes, that would make it so much easier for me. If I could just hear his voice . . .'

'Have you tried to get his number?'

'No. I've thought about it – but no.'

'What about going to Dubai? Have you thought about following him to the Middle East?'

'Yes. I have.'

'You're still here though . . .'

'Yes,' she said. 'I'm still here.' Then she sighed, a colossal expulsion of air that created an illusion of shrinkage. Her shoulders curved inwards and her knees ascended slightly as her heels left the ground. This diminution, this closing in, was strongly suggestive of the foetal position. Her hands became fists, held tightly against her stomach. Then she added: 'I know . . . I know.' Her eyes were glistening.

What did she know?

She had allowed herself to contemplate the possibility that Daman Verma didn't love her, that their love was not fated, and that they would never be together. She had looked into the abyss and the pain she felt was devastating. 'I know . . . I know.' That's all she said. I can still re-create the sound in my mind, even down to the acoustic properties of the room we were sitting in: hesitant, slightly hoarse – a double cadence – full of sadness and resignation. I had told Megan not to over-interpret, but the register of her voice, her posture, the trembling light in her eyes, gave expression to her thoughts with pitiful eloquence. It was plain what she was thinking, and her grief was palpable.

Falling in love *is* painful. Most of us know what it's like – the need, the desperation, the longing. And when we aren't loved in return, the anguish can be unbearable. Time heals, but it isn't time that gives us the courage and strength to carry on. We carry on because of hope; hope informed by experience and observation. We learn, either directly or indirectly, that love is not always reciprocated, overtures are rejected, and relationships full of early promise fail, but we also come to appreciate that opportunities to find love will inevitably come again.

Megan had found the love of her life. She was devoted to him, and her devotion was equal to all the well-worn, extravagant metaphors of poetry and song. She was as constant as the sun, the moon and the northern star. There would never be a transfer of affection. So there was no hope, no future. The distress that most of us might have to tolerate for months or years she was going to have to tolerate for the rest of her life. Imagine it. Remember what it feels like to be desperately and unhappily in love – and now imagine those same agonies sustained without respite, in perpetuity.

'It's so unfair,' Megan whispered.

'Yes,' I agreed. 'It is . . .'

The tears tumbled down her cheeks and splashed on her skirt. I pushed the box of tissues towards her. She didn't notice my wholly inadequate gesture. She was too far gone – and I was humbled by the sheer magnitude of her agony.

What are the causes of de Clérambault's syndrome? The most accurate and intellectually honest answer to this question is also probably the least satisfactory. No one really knows. It has been attributed to neurotransmitter imbalances, but the medication employed to correct those imbalances is rarely effective. Dopamine might have a role to play – but Megan's medication, which worked by blocking dopamine receptors in the brain, had no effect on her mood, thinking or behaviour. Most patients report a dulling of emotion but the underlying fixation persists.

Another possibility is abnormal electrical activity in the temporal lobes – particularly the right temporal lobe. De Clérambault's syndrome and temporal lobe epilepsy (TLE) share some common features: intensification of emotions, altered sexual interest and transcendent episodes. When the latter occurs the patient is sometimes said to be suffering from 'Dostoyevsky epilepsy' because the famous writer was prone to ecstatic seizures. Some individuals with TLE have insisted that strangers have fallen in love with them – although this is extremely unusual.

Psychoanalysts have implicated sexual ambivalence. By choosing an unattainable lover, the sufferer is able to avoid intimacy. Once again, the theory isn't compelling, particularly with respect to cases like Megan. She had enjoyed a normal sex life before she encountered Daman Verma. She hadn't been avoiding intimacy at all. Another theory suggests that women afflicted with de Clérambault's have unaffectionate fathers. That, of course, is true of many women – but they don't all go on to develop the condition.

De Clérambault's syndrome is difficult to treat. The prognosis is poor, and the illness usually has a chronic course. A combination of medication and enforced separation is supposed to be the most effective treatment, but Megan had taken Pimozide and hadn't seen Daman Verma for six months and she still yearned to be with him.

One day, I asked Megan if she thought we were making any progress. 'Yes,' she said. 'It's helpful . . . talking.'

I flattered myself we were getting somewhere. But I was very much mistaken.

All things being equal, we tend to pair up with an individual of a similar type to us – particularly so with respect to attractiveness. If you want to get an idea of how good looking you are, don't look in the mirror: take a long hard look at your partner instead. In evolutionary terms, beauty is only one of many fitness indicators but it may be the most important. Everyone wants to find an attractive mate and few are willing to couple with anyone less attractive than themselves. Beautiful people pair with other beautiful people and the not-so-well-endowed must do their best in a depleted market place where they will continue to resist trading down. These imperatives create a hierarchy in which the vast majority of couples have sorted themselves into closely matched pairs. Evolutionary theorists call it assortative mating. Exceptions are relatively rare, and when they do occur they are often attributable to the influence of wealth (another fitness indicator), which tends to facilitate relationships between richer, older men and desirable younger women.

I wondered what Megan's husband Philip was like. So I asked to see him.

Philip was the same age as Megan and roughly the same build. He had the same coloured hair and was, at most, only one to two inches taller. He dressed in the same way: respectably casual – a pale blue shirt, a dark blue jumper, grey flannel trousers with a neat crease, and polished Oxford brogues – but not so casual as to preclude dropping into the office. His manner was affable and pleasant. I recognised his deferential, self-conscious smile, because it was an exact copy of Megan's. It was easy to imagine them, as a couple, before the catastrophic entry of Daman Verma into their lives, hand in hand and ideal companions.

'The last few years must have been very hard for you,' I said.

'Yes,' he replied. 'It has been quite difficult.'

Here was a man with a gift for understatement.

We talked a little about the nature of his relationship with Megan and how things had changed.

'I suppose since Daman went to Dubai, things have been better.' He used the dentist's first name, just like his wife. 'I mean, I don't have to worry about where she is – or what she's doing. She's back at work now and she comes straight back home. They've been very good, the people who she works for. The head clerk particularly. His daughter suffers from depression and he was very understanding.'

'Do they know what happened?'

'Well . . . not exactly.' He hurried on, not wishing to dwell upon the fictions he had employed to minimise embarrassment. It was sad that he'd had to lie, and, of course, an indictment of our society. Even when a sympathetic reception was guaranteed, Philip couldn't tell the truth. It was still too shameful and humiliating.

'On the face it, it's like everything's normal again. We chat – go to the cinema – go for walks. We went to Cornwall for a few weeks last August and had a really nice time.'

'Are you still . . . close?'

'Yes, I think we are.'

I wanted to know *how* close. 'Are you still . . . intimate?'

'Intimate? What – having sex?' I nodded. 'Yes,' he continued. 'Yes we are. It's so strange.' He suddenly looked bewildered, boyish. 'Nothing's changed – but everything's different.'

'How do you mean?'

'My wife is *there* – but not there. It's *her* – but not her.'

His words reminded me of a clinical phenomenon known as the Capgras Delusion. The sufferer believes that someone to whom they are closely related has been replaced by an identical imposter.

'I know she's thinking about him all the time,' Philip continued. 'I mean – she's probably even thinking about him when, you know, we're in bed.'

'You think she's having sexual fantasies about him while you're—'

Philip cut in, preventing the sentence from reaching its obvious and explicit conclusion. 'No, no.' He took a deep breath, composed himself and added, 'Well, I can't say for sure of course. I realise that. Maybe she does think of him while we're making love. But I don't think so.' Philip believed that Megan's feelings for Verma had become more abstract – more elevated. And he had good reason.

'Has Megan told you about her . . .' His last syllable extended equivocally. He scratched his head, as if he'd been presented with an intractable mathematical problem. 'I don't know what to call it really. I suppose it's like a shrine.'

'What?' I sat up, surprised. 'No, she hasn't told me.'

'It's a box – an ordinary storage box – which she keeps in the bedroom, covered with a white cloth. Inside the box are things she's collected that have some connection with Daman.'

'Such as . . . ?'

'He was in the papers once. He'd attended some big fundraising event for a charity and his photograph had been taken. He's dressed up – in a tux – standing next to an MP and a TV personality. It all looks quite glitzy. Megan cut the article out of the newspaper and kept it. She also has his old business card, an information pamphlet that she picked up at his clinic and her appointment letters. And there are a few other things. A pen, a paper clip . . . I can only imagine that they're things that he touched. She must have stolen them.'

'What does she do with these things?'

'She takes them out from time to time.'

'In front of you?'

'No. She used to, but not now. She used to sit next to the box and close her eyes. It was as though she was – I don't know – praying.'

'How do you feel about this . . . this shrine?'

Philip looked discomfited by my question. 'It's just something I have to put up with, isn't it?' The look of boyish bewilderment returned.

'No. Not necessarily. You could say something.'

'Could I?'

'Yes. You could object.'

He shook his head. 'I couldn't force her to throw those things away. It would be crushing. Why would I do that? Why would I *want* to do that?'

I was touched by his compassion. Ordinary, non-pathological love can also be very extraordinary.

The next time I saw Megan, I asked her about her shrine.

'It's the closest I'll ever get to Daman now – physically, I mean.' Her addendum was a telling qualification. She still believed that the great distance that separated her from Verma could be bridged by non-physical means.

'How often do you look at those things?' I asked.

'Not very often, but it helps – knowing they're there.'

'How do you think Philip feels about you keeping these . . . mementoes?'

'He doesn't mind.'

'Are you sure?'

'Yes. He doesn't mind. And it's not doing any harm.'

'Maybe if you could let go of those things, it would help you to move on.'

A shadow, something like fear, darkened her face. 'It's not doing any harm. And Philip doesn't mind – really he doesn't.' The note of barely concealed panic rang out all too clearly.

Fictional representations of psychotherapy are very misleading. A heroic clinician is summoned to treat an unreachable patient whose symptoms defy understanding. With considerable difficulty, demanding a combination of insightful brilliance and guile, and against all odds, a relationship is established. Dark discoveries are made by the excavation of unconscious memories and the mystery is finally solved. All the pieces of the complex puzzle fit neatly together and the patient is restored to perfect health. Exit hero therapist – cue music and titles.

The reality of psychotherapy is very different. It's actually quite messy and rarely progresses along the satisfying lines of a fictional narrative. There are blind alleys and false turns, periods of stasis and frustration – doubts about whether one is addressing the problem in the right way. Even when attempting to treat a specific anxiety with a straightforward method such as 'exposure' – which involves persuading patients to confront their fears directly – something can happen which recommends adopting an entirely different approach. I was once conducting an exposure session with a woman who suffered from a horror of door handles because she was afraid of being contaminated. As she anxiously reached out to touch the handle of my office door, she remembered another door handle, the one that had rattled ominously before her father entered her bedroom to sexually abuse her when she was a child. Needless to say, we abandoned exposure and spoke about these memories instead. Theoretically dense therapies, such as psychoanalysis, can easily feel unnavigable; all those memories, dreams and interpretations. The unconscious isn't always cooperative and it is possible to dig deep into someone's psyche and uncover nothing of therapeutic value.

The pieces of Megan's puzzle didn't fit together neatly. There were no dark discoveries and I couldn't discern any pleasing, explanatory connections. A staunch biological psychiatrist would probably suggest that this is because de Clérambault's syndrome is a psychotic illness and best explained by chemical imbalances in the brain. I was looking for things that weren't there or were merely incidental. The fact that Megan's medication didn't work doesn't compromise this argument. Perhaps we just need better drugs.

I can't offer a psychological explanation, but I can offer an observation – a kind of contextualisation that has certain implications for how we view patients like Megan.

The more I thought about Megan, the more I was struck by the correspondences between her so-called illness and the behavioural and emotional correlates of romantic love. Her abnormality was quantitative rather than qualitative. She was experiencing the same things that we all experience when we are smitten, only greatly magnified. Even her delusional thinking was, in a sense, normal, because romantic love is often very irrational – love at first sight, ascribing chance meetings to destiny, oceanic feelings and powerful affinities that can transcend time and space are all commonplace. Most love-struck individuals engage in subtle forms of stalking – for example, loitering in places where they are likely to encounter the person who they've fallen in love with. Even Megan's shrine can be viewed as just an exaggerated version of the photographs and sentimental objects that couples often retain to memorialise their love; relics and talismans that contain residual energies released at the time of a first meeting, dinner, or kiss. The only feature of Megan's illness that marked a qualitative departure from normality was her absolute conviction that Daman Verma was also smitten, a conviction made even more conspicuous by its survival regardless of overwhelming evidence to the contrary. Other than this delusion of reciprocity, Megan's psychopathological love was simply romantic love writ large: not abnormal, as such, but supernormal.

It is as if the neural circuitry that serves romantic attachment – the same neural circuitry laid down by natural selection and shared by all humans – suddenly became hyperactive. What this suggests is that what happened to Megan could also happen to any of us. And if you have ever fallen in love, you will, no doubt, have edged closer towards Megan's location along a continuum. Many – none of whom are ever given a psychiatric diagnosis – travel most of the distance.

Psychologists make a distinction between problem-focused and emotion-focused coping. The former is what we do when a problem is soluble. If you have to sit a difficult exam you can always do more revision. Some problems, however, are insoluble – like bereavement, for example – and then the only option becomes changing one's response to the problem. This is, of course, a major undertaking – but at least it is theoretically possible.

Did I help Megan? There was no solution to the problem of Megan's de Clérambault's syndrome – she was incurable – but she did modify her response to the problem. She came to accept that she would have to live her life separated from Verma, and to the best of my knowledge she never attempted to follow him to Dubai; however, she still loved him – and would love him forever.

Although I saw Megan a long time ago, I still think about her. I imagine her, surreptitiously climbing the stairs to her suburban bedroom, entering and closing the door. I imagine her sitting at her shrine and removing one of the sacred objects from inside. I imagine her, closing her eyes and communing with a man who has probably forgotten that she exists by now.

Chapter 2

The Haunted Bedroom: *Ageless passion*

A grey, overcast day in autumn. I looked out of the window through streaming rainwater at a grim prospect: a narrow paved pathway, hemmed in by temporary huts, leading towards a cliff face of soulless 1960s architecture – a no-man's land between a research institute and a psychiatric hospital. The people who traversed this desolate corridor were mostly psychiatrists and nurses, but occasionally I'd see a stray patient. One of them was a black woman who always covered her face with white make-up because she believed she was an angel. She must have thought it necessary to have white skin in order to qualify as a member of the angelic host. Her appearance was actually quite disturbing, but whenever I encountered her on the street she gave me a friendly smile. It wasn't always easy to distinguish staff from patients. Another individual I frequently observed through the window was a donnish gentleman in his sixties, dressed in a crumpled polyester suit and incongruous trainers. He was always sprinting and jogged on the spot even when travelling between floors in a lift. Over a period of several years, I never saw him stationary. I later learned that this mercurial eccentric was not only a renowned physiologist, but also a musicologist, composer, former member of the Ratio Club (which included Alan Turing as a member), and the inventor of an electronic wind instrument known as the logical bassoon. He also enjoyed some notoriety on the conference circuit. Once, he injected his penis with an impotence cure and encouraged delegates to admire the strength of his erection. The propriety of his behaviour was never questioned. Obviously, these were different times.

My office was in an Edwardian terraced house next to the hospital grounds. It was used as an outpatient clinic. I was told – before starting work there – that the building had been repeatedly inspected by the local council and condemned. The house always escaped demolition because of the pressing need for more space. I laughed, assuming that the story must be apocryphal, but one morning I opened the door to the consulting room and half of the ceiling had fallen in. There was a massive hole through which I could see floorboards and water pipes. Everything was covered in fragments of plaster and dust.

The house was in a terrible state of dilapidation. Paint was peeling off the woodwork and colonies of black mould climbed up the walls. The furniture was the kind you might find in a junk shop. I can vividly recall a penniless patient (a man who lived in one of the local housing estates) asking if I'd accept a charitable donation.

A gust of wind rattled the window pane, and a nurse, pulling the lapels of her coat over her head to make a cowl, hurried along the path. The bell rang and I went to let Mavis in. We hadn't met before, but I was familiar with her history from the referral letter: a working-class woman who had lived in the same underprivileged borough all of her life. She was in her early seventies and very depressed; the cause of her depression was the death of her husband. He had suffered a fatal heart attack a year earlier.

When individuals have significant psychological problems after bereavement (lasting more than twelve months) they are said to be suffering from a complicated grief reaction or Persistent Complex Bereavement Disorder. I find the idea of conceptualising prolonged grief as a form of abnormality questionable. People vary in temperament and resilience and come to terms with loss at different rates. Some never adjust. The fact that such an appalling trauma might cause long-term distress is hardly surprising. I'm inclined to ascribe protracted grief to the human condition. Diagnosis seems a rather arcane consideration in this context.

I opened the door and faced a small, slightly overweight woman holding an umbrella over her head. The colour of her hair matched the sky and her expression suggested vacancy. When people get very depressed, they don't look sad but exhausted. It's as though they've progressed beyond sadness and have resumed their existence on another, unreachable plane. Mavis looked emotionally numb, but numbness implies anaesthesia and that would be misleading. The numbness of depression is simply pain in another form – like water becoming ice when the temperature drops. Dante knew what he was doing when he characterised the ninth and lowest circle of hell as a frozen waste.

'Come in,' I said.

'What should I do with this?' She indicated her umbrella.

'You can leave it here in the hall to dry if you like.' I tapped the radiator.

She stepped out of the rain, placed her umbrella on the floor and followed me to the consulting room. She didn't register the shabby state of her surroundings – the small, circular cigarette burns in the carpet, the general atmosphere of decrepitude. She sat down on a worn armchair with noisy springs and faced me with her knees pressed together. She was wearing a pleated blouse, a loose cardigan, a dark skirt and grey woollen tights. I made some introductory remarks, summarised her referral letter, and checked that she understood why she'd been advised to see me.

'I'm not doing very well. That's what *he* said – Dr Patel.' Her voice sounded querulous. 'You know, since George died. He – Dr Patel – reckoned I should talk to someone. He said it might help.'

Even though psychotherapy can be a challenging occupation and many patients don't get better, there is always a possibility, however slim, that treatment will be successful. A woman with agoraphobia might be persuaded to leave her house or a man with obsessive-compulsive disorder might learn to resist his compulsions. But death is irreversible. As a psychotherapist undertaking bereavement counselling you can only ever tinker at the edges. Talking to a psychotherapist doesn't bring back the dead.

Mavis was difficult to engage with. When she answered

questions she tended to favour monosyllables. Even so, I tried to keep things going. I kept prompting her, sometimes with words, sometimes with expressions or gestures, until our exchanges achieved a kind of rhythm, enough momentum to carry us forward.

She had left school without gaining any qualifications and married young. Her husband, George, had been a postman. Two years after they were married, their son Terry was born. Mavis had been a housewife all her life and had never been tempted to get a job outside the house. When Terry left school he worked in a factory. Eventually he became a foreman. He was now in his forties and still lived at home. I asked Mavis if her son was in a relationship.

'No – he's not much of a ladies' man.'

'Oh?'

'He likes his money.'

'I'm sorry?'

'He doesn't want to share it with anyone.'

'Does he share any of it with you?' Ageing mothers make excellent and very cheap housekeepers.

She answered loyally: 'He pays his way.'

I wasn't convinced. 'Has he ever had a girlfriend?'

'When he was younger, but he hasn't for a while now.' Terry didn't sound like a very generous-spirited man. He was more interested in cars than people. 'Always outside, he is: fixing his Mini up. It's his hobby – customising.'

I wanted to know more about George.

'He was never a talker. He'd come home, have his dinner, and then we'd watch the telly.'

'What sort of things did you do together?'

'Do?'

'Yes . . . when you were together.'

'Well, we didn't go out much, if that's what you mean.' She appeared to be surprised by my question, as though the idea of doing things together as a couple was unheard of – foreign or suspicious. 'Sometimes we'd do a bit of shopping on Saturdays. We'd go down to the market. But not often – there was no need. I did the shopping in the week.'

I asked her about their social life.

'George didn't have many friends. He'd go for a drink once in a while – but that was it really.'

'What about you?'

'Me?' She shook her head. 'I had my husband . . . '

Mavis was lonely. She thought about George every day – all the time, in fact. His absence had opened a chasm in her being, a frigid void. She missed him and she missed him terribly. Yet, when Mavis spoke about George, it was difficult to understand what she was missing exactly. I didn't get any sense of their life together; there were no fond reminiscences or anecdotes. And there was something very odd about how Mavis talked about their son. Children cheat death. Our expressions and mannerisms survive in our children. If a father dies, a mother might still find solace in the reproduction of his smile on her son's face. Yet, Mavis talked about Terry as if he were a lodger.

Ostensibly, Mavis was coping, doing the same things she'd always done: the housework, the cooking, Terry's washing and ironing. But she was functioning like an automaton. I asked her if she was still able to get pleasure from anything. 'Food,' she replied. 'I still treat myself now and again. Sponge fingers, evaporated milk – fruit cocktail.'

Undertaking psychotherapy with older patients, particularly those who have not been fortunate enough to receive an extended education, can be testing. They often find it difficult to express their feelings – having always been told not to. They can be inflexible and unable to grasp abstract ideas. Mavis was hard to help for all these reasons. But in addition, I sensed there was something else, something important that I hadn't identified.

While discussing her loneliness, I asked Mavis what she missed most about George. I had chosen to ask a direct question in order to elicit a direct answer.

She looked at me through the smudged lenses of her spectacles and replied without hesitation, 'The sex.'

I have to admit, I wasn't expecting that.

We think of sex as a drive, powered by hormones; however, sexual motivation is much more complex and subtle. Hormones are only part of the explanation. Although there is an association between levels of testosterone and desire, it is possible to have high levels of the former and little or no desire. Similarly, removal of the testes – where most of the testosterone in the male body is produced – does not always lead to loss of sexual interest.

We are motivated to have sex when certain circuits in the brain become active in response to sexual thoughts, images, or external stimuli that we find arousing. These circuits are sensitised by hormones and signals that are transmitted from the genitals.

Contemporary psychologists explain sexual desire within a framework called incentive motivation theory. We are not pushed towards sexual objects by a drive; rather we are pulled towards them. We are attracted by incentives. The value of sexual incentives is determined by the outcome of our past sexual experiences. Pleasurable outcomes will raise the value of incentives, whereas unsatisfactory outcomes will lower them.

Many people enjoy sex when they are old. But desire dwindles with age – even more so over the course of a fifty-year marriage. Our bodies change and as a consequence so do our needs and appetites. The psychologist Robert Sternberg has suggested that love, when it meets the standard of our cultural ideal, is composed of three elements: intimacy (or closeness), passion (which is mostly sexual) and commitment. Sternberg calls this 'consummate love'. These three elements are not always present or equally represented and they can be combined in different ways to produce less enduring or satisfying forms of love. For example, passion alone, without intimacy and commitment, produces a highly unstable 'infatuation', whereas intimacy and commitment, without passion, produces 'companionate love', which has the character of an affectionate, long-term friendship.

Assuming that commitment holds steady, as years of married life accumulate there is usually a shift of emphasis, from passion to intimacy. Sex isn't such an urgent necessity and relationships become more companionable and emotionally rewarding.

The first major reduction of passion occurs after three or four years of marriage and many relationships fail around this time; indeed this is when divorce statistics peak. The reason for this is probably evolutionary, three to four years being the optimal time to reproduce and ensure the survival of offspring in the ancestral environment.

When men settle and start a family, their levels of testosterone drop and will continue to drop unless a new relationship is formed. Although testosterone is often described as the male sex hormone, it is also associated with sexual activity in women, whose levels of testosterone follow much the same pattern as men with respect to marriage and procreation. Women also experience an additional loss of libido in later life which is linked to post-menopausal reductions in testosterone.

Sex is precious, because for most of us, it is time limited. Even if couples continue to have sex into their eighties, it can never be as vital as the sex they had when they were eighteen. Lack of stamina, blunted senses, failing health and low testosterone inevitably make the experience less physically intense. There must be few people who, languishing on their deathbed, wish they'd had less sex when they were young. Sex was at the heart of Mavis and George's relationship. Indeed, it may have been the entirety of their relationship.

'We used to do it all the time,' Mavis confessed, clearly bemused by the improbable strength and longevity of her libido. Her voice didn't soften when she said this. There wasn't a hint of a smile. I wondered whether her feelings about sex were complicated by guilt. Many women of her generation were taught to view recreational sex as morally suspect. My efforts to explore this possibility were promptly repelled with a typically concrete response. 'No, I never felt guilty about it. Why should I? We were married.'

A marriage based almost entirely on sex shouldn't last. It should begin to fail after a few years. Sternberg's theory of love is known as the triangular theory, because 'consummate love' requires all three – intimacy, passion and commitment – to be equally balanced. Like a tripod, 'consummate love' stands on three legs. If you remove just one of these legs, it falls over.

Of course, a tripod can still stand – up to a point – if one of the three legs is shorter than the other two. This is precisely what happens when passion cools. Marriages become inherently less stable but the weight-bearing distribution of passion, intimacy and commitment is sufficient to prevent the tripod from toppling.

Why did Mavis and George's relationship last? They didn't talk very much and didn't have common interests. They were committed, certainly, but Sternberg calls the combination of commitment and passion, in the absence of intimacy, 'fatuous love'. It is without real substance and largely irrational. Why would anyone commit themselves to a person they didn't really know? For commitment to be meaningful it must be preceded by amity. Moreover, fatuous love is always doomed because passion declines leaving nothing but empty commitment. When this happens, couples only stay together out of a sense of duty – and not for very long.

None of this applied to Mavis and George. Their passion didn't diminish as the decades passed and the sex was good enough to keep them committed to each other for fifty years. They didn't need to talk.

Mavis still yearned for her husband's touch, the feel of his body against hers, physical contact; and this yearning became so great that it was like a cry, a calling out – or something even more powerful – a summoning.

'I can still feel him – you know, like he's still around.'

It was an important disclosure. I wanted her to continue but detected signs of internal struggle. Neither prolonging the silence nor posing a question seemed right, so I reflected her last statement back.

'You feel like he's still around . . .'

Mavis nodded. 'When I'm in bed.' She paused and looked at me with a peculiar, concentrated expression. 'One morning I woke up and I saw him – standing by the wardrobe. You know . . . a ghost.'

'What did you do?'

'"George," I said. "George." But he faded away.'

Do ghosts exist? Of course they do. There are simply too many reports of ghostly sightings to doubt their existence.

Two theories predominate. The first is that ghosts are spirits returning to the world after death and the second is that ghosts are a psychological phenomenon. Today, we tend to favour the latter. However, just because something is psychological doesn't mean that it isn't real. Your memories, for example, are every bit as real as a rock, a tree, or the sun. Although a brain scientist might argue disparagingly that memories are a mere by-product of underlying biological processes, epiphenomena, this doesn't make them any less real. They are simply real in a different way.

It's often the case that fiction identifies subtle truths more readily than science. This is particularly true with respect to ghosts.

The first truly psychological ghost story was Henry James's *The Turn of the Screw*, published in 1898. The plot is very simple. A governess becomes convinced that two ghosts (former employees of the house and lovers) are wielding a malign influence over the children in her care. She challenges their supernatural authority and a tragedy ensues.

What makes it psychological is the way the story is told. We can't help but wonder if the ghosts are supernatural or imagined. We are tempted to engage in a little amateur psychoanalysis. Is there, perhaps, some connection between the ghosts and the prim governess's repressed sexual feelings? In this respect, James prefigures Freud, who suggested that supernatural occurrences represent the return of the repressed. Ghosts are projections from the unconscious.

The psychological ghost story is extremely powerful, even to modern readers, because it does not depend on credulity for its effect. We are not asked to believe in disembodied, vengeful spirits, but in the self-evident truth of the human mind. Thus, the ghosts are real – as real as our memories and our forbidden wishes.

Mavis's forbidden wish was all too apparent. She wanted sex with her husband and her unconscious wasn't recognising death as an obstacle.

It transpired that Mavis had encountered George's ghost more than once. Over a period of several weeks, she became more comfortable discussing the subject and she revealed that she had seen George four or five times in her bedroom and twice in public.

'I was sitting in the park and I saw him standing under a tree.'

'Did he look . . . real?'

'Yes, just like he did when he was alive. He was wearing his mac.'

'What did you do?'

'I was getting my things together, getting ready to go over and talk to him, but when I looked up he'd gone.'

Apparitions of this kind tend to be elusive: a blink, a small head movement, or a subtle change of light – like the sun appearing from behind a cloud – can easily dispel a ghost.

'Where else have you seen George?' I asked.

'On the high street, but I lost him in the crowd.'

I was inclined to count this latter sighting as a case of mistaken identity. 'Okay . . .' I nodded. 'Okay.'

Hallucinations are defined as perceptions that arise in the absence of external stimuli. The most common forms of hallucination are auditory and visual, but equivalents occur across all sensory modalities. For hundreds of years, hallucinations were regarded as a cardinal symptom of madness, a reliable marker of abnormality. But this isn't correct and probably never has been.

What we think of as objective reality is a kind of compromise, the result of external stimuli impinging on the senses followed by interpretations. Eyes have blind spots and change position every second, peripheral vision is extremely poor, and the image that arrives on the retina is small and indistinct. We should see a foggy, shaking world with dissolving edges and parts missing. But instead, our picture of the world is complete, stable, panoramic and sharply defined. This is because a great deal of unconscious editing has taken place before visual information enters awareness. The brain fills in gaps, compensates for movement and makes educated guesses. This editing process is biased by our expectations, motivations and desires. A new mother, for example, will constantly misinterpret background noise as her baby crying. Even when there is no noise at all she will still experience the occasional false alarm. She will cock her head to one side and say, 'Did you hear that?'

The cognitive psychologist Roger Shepard has said that perception is 'externally guided hallucination' and hallucination is 'internally simulated perception'. In other words, reality isn't entirely authentic and hallucinations aren't complete fabrications.

Approximately 5 per cent of adults hallucinate and never seek medical help. They just accept that they are hallucinating and carry on as normal. Moreover, a third of Americans claim to have seen angels. This may sound like an exaggerated figure, but actually it is entirely consistent with the fact that roughly a third of children have imaginary friends.

The type of hallucination that Mavis reported is extremely common and has been given a precise name – PBHE or Post Bereavement Hallucinatory Experience. Some studies have found that as many as 80 per cent of bereaved individuals report PBHEs. This suggests that these experiences are a normal rather than an abnormal phenomenon. If you are predeceased by your partner, you will probably see them again before your own demise.

People don't talk about their PBHEs very much. Perhaps this is because the experience of seeing one's wife or husband after they've died is so strange and deeply personal that it's hard to know where to begin. How does one broach such a topic? Or perhaps it's simply that people worry about attracting a psychiatric diagnosis.

When my mother was dying in her hospital bed, she was no longer lucid. When she opened her eyes, nothing registered. The only sound she made was an intermittent 'Ow, ow, ow' as if someone were prodding her repeatedly with a pointed stick. It was distressing to watch and she continued a whole night and into the early hours of the morning. She didn't appear to be in great pain, but she did look extremely annoyed.

My mother's best friend was sitting next to me. 'She knew what to expect.' Resting a solicitous hand on my arm, the friend continued. 'She's had plenty of time to prepare herself.'

'I don't think any of us are ready to die,' I replied. 'Not really.'

'No,' my mother's friend insisted. 'She was. Because of your dad . . . '

My father had died a decade earlier.

'I'm sorry?'

'She could feel his presence. And it got stronger and stronger. It's so strong – she used to say. Sometimes I can feel him in the house and I call out his name. Other times it's just like he's sitting beside me. She knew. It was like he'd come to get her.'

Over a period of ten years my mother hadn't said a word to me about my father's spectral visitations. I can only assume that this was because I didn't share her religious convictions. By the time the friend had disclosed my mother's secret, my mother had already spoken her last words. She never spoke again, and the following afternoon she passed away. I'm still curious.

'Do you think it's him – really?'

Mavis's spectacles had slipped and she pushed them back up her nose. 'Yes.'

'What do you think it means? George returning . . .'

'I don't know. Maybe he's missing me too.'

'Are you religious in any way?'

'No. I'm not a churchgoer – never have been. George neither.'

'Do you believe in an afterlife now?'

It crossed my mind that she might find some consolation in faith – or at least find companionship and support as part of a congregation.

'I'm not sure. I don't know what I believe.' One of Freud's great theoretical contributions was to suggest that contradictory beliefs can co-exist in the unconscious. In fact, human beings are even more perverse. We can be knowingly inconsistent. For Mavis, her dead husband's appearances in the bedroom didn't seem to have any spiritual implications.

'All right.' I made some unnecessary notes, while I thought of my next question. 'Has George ever spoken to you?'

'No. He just appears . . . and then fades away.'

'Are there any alternative explanations for these experiences?' She didn't grasp what I was getting at. 'Is it possible that what you're seeing isn't really George, but rather, a kind of illusion?'

I wasn't seeking to challenge her. I was simply trying to get a better understanding of how she was appraising the situation. She was peculiarly unreflective and passive – disinclined to question and analyse what was happening to her.

She answered plainly: 'No.'

'You don't think it's possible that maybe you're missing George so much that your mind is playing tricks?'

She pressed her lips together and frowned before repeating: 'No.'

'Mavis,' I put my pen down and leaned forward. 'How do you feel when you see George?'

'I'm not frightened . . . '

I could see that my questions were confusing her so I didn't persevere. Besides, her hallucinations didn't seem to be doing her any harm. Perhaps PBHEs are not so much a symptom of poor adjustment, but rather the product of a protective and adaptive stress response. They must, at some level, make death seem less absolute and ease loneliness.

Mavis came to see me for about ten sessions in total. I did exactly what I was asked to do – bereavement counselling and some Cognitive Behavioural Therapy (CBT) for her depression. The latter involved encouraging her to perform simple experiments to test some of her rigid and unhelpful beliefs. These beliefs mostly concerned what she thought she could and couldn't do. She didn't, for example, feel that she could cope with social situations.

While I was seeing her, she became a little more active and even started to attend a social club run by a mental health charity. But the gains she made were relatively small, because in reality, she didn't want to make friends, improve the quality of her life, or even feel happier. What she wanted, more than anything else, was sex. Ghostly visitations are usually invested with spiritual significance – they confirm that the soul survives death and promise a heavenly reunion. For Mavis, the opposite was true. Mavis didn't – or couldn't – think in this way. Her husband's ghost, shaped by carnal desires, offered no prospect of transcendence, only deeper and deeper entrenchment in the flesh.

The psychiatrist Elisabeth Kübler-Ross identified five stages of grief: denial, anger, bargaining, depression and, finally, acceptance. Even though her work has been extremely influential, there isn't a great deal of evidence to support the notion that grief can be neatly partitioned. Loss is experienced uniquely and can have different meanings and different consequences – depending on the individual. There is no one-size-fits-all approach to bereavement and no right way to grieve.

It was raining again when I saw Mavis for the last time. She picked up her umbrella in the hallway and I opened the door for her.

'Goodbye,' I said. 'And if you ever want to talk again, all you have to do is telephone the psychology department.'

We shook hands. She didn't smile. Looking up at the sky, she pressed a button on the handle of her umbrella and the black canopy expanded. She descended the stairs and walked towards the pub on the corner. I thought she might glance back – but she continued walking and disappeared from view.

I went back inside and stood by the window overlooking the bleak space between the hospital and the research institute. The eccentric physiologist – in his polyester suit and trainers – sprinted through the deluge. No one else appeared and the paving stones began to glisten as the lamps came on. I thought about Mavis for some time.

Many people complain about sexless marriages. But Mavis's relationship with George was based *primarily* on sex. They should have been beached by the retreating tide of testosterone – stranded together with nothing to talk about – two strangers, waking up each morning next to each other before going their separate ways. Instead, they were like lotus-eaters, inhabiting a bedroom paradise of sensual, indolent pleasures – emerging only to face a dull reality that, at best, could only offer them the poor consolation of tea and sponge fingers. Sex shouldn't have retained its adhesive properties. They should have fallen out of love, because sex isn't love – only a part of it. But Mavis and George had demonstrated that if passion doesn't flag or die, the other components of love become expendable.

Why was that surprising? It shouldn't have been really. In the first throes of a relationship, when desire is at its strongest, couples are bound together more closely by sex than by conversation. Desire is more powerful than liking.

I remembered a famous Woody Allen quote: 'Sex without love is a meaningless experience, but as far as meaningless experiences go it's pretty damned good.'

Mavis and George's indestructible passion must have had

consequences for their poor son. Originally, I'd thought of Terry as mean and selfish, but I found myself feeling sorry for him. We all want our parents to love each other – but not too much. Parents who love each other passionately – and keep on loving each other passionately – produce orphans.

Was that why Terry was still living at home in his forties? Had he been waiting all his life to be loved? Well, now that his father was gone, perhaps there was a chance.

The grim reaper always throws a sly glance as he exits the stage – hinting that his secret gift is renewal.

Chapter 3

The Woman Who Wasn't There: *Suspicion and destructive love*

My patient hadn't materialised.

I wrote the date in the margin of his notes followed by the letters DNA – 'did not attend'. Clinicians often use abbreviations: SSRI (Selective Serotonin Reuptake Inhibitor), CBT (Cognitive Behavioural Therapy), PTSD (Post Traumatic Stress Disorder). The main reason why abbreviations are so popular is because much of the clinical vocabulary involves long compound terms and using them in conversation quickly becomes tiresome. Abbreviations are a form of exclusive communication between health professionals and they can acquire the properties of slang. I once worked in the children's department of a hospital where I learned the meaning of FLK – funny-looking kid. Sometimes, it isn't always possible to attach a diagnosis to a child but there's *something* about them that nags at the back of the mind and suggests things aren't quite right.

This elusive *something* is almost always reducible to facial characteristics and can signify neurodevelopment problems. Peculiarities of gesture and gait are also significant markers. Although FLK is an insensitive abbreviation it is by no means the worst. Some stressed doctors use LLS: looks like shit.

I put the case notes of my DNA aside and picked up another folder. On opening it, I discovered a terse and uninformative referral letter. My next patient was a woman in her late thirties with relationship problems. After reading the single paragraph I spent forty minutes either flicking through an academic journal or pacing around the room. A high number of patients fail to attend mental health appointments, which means a psychotherapist spends many hours hanging around, drumming fingers, checking wall clocks and staring out of windows. It feels a little like having been stood up – only it happens every day. The telephone rang and my secretary announced the arrival of my eleven o' clock patient.

Anita was a striking woman – a tall, leggy blonde with eyes an astonishing shade of violet. She was dressed casually in jeans and a jumper but managed to look very elegant. It was as though she'd grabbed whatever first came to hand and by chance the combination happened to be perfect. I learned later that she was an interior designer.

'So ...' I said, opening her folder. 'I understand that you've been having relationship difficulties.'

'Yes.' She looked as if she might say more but her expression changed and the hesitation became silence.

'What's your partner's name?'

'Greg.'

'And how long have you been together?'

'About a year.'

They had met at a dinner party of a mutual friend. Greg was a games programmer who had started his own company. It was a lucrative business and one of his games had won an award. 'Computer games aren't really my thing,' Anita said, wrinkling her nose. 'I thought he was going to be a nerd, but we got talking and something clicked. The chemistry was right.'

People often employ the word chemistry when they can't explain mutual attraction. Goethe's extraordinary novel *Elective Affinities*, which was published in 1809, explores the idea that romantic attachments might obey the same laws that predict the formation of chemical bonds. Men and women are certainly responsive to each other's bodily secretions, which autograph the air with unique molecular signatures. When inhaled, these molecules can promote hormonal changes associated with sexual readiness – even when there is no conscious detection of smell. In the sixteenth century, women placed peeled apples under their armpits to impregnate the flesh with perspiration. These saturated fruits would then be presented as a gift to prospective lovers who subsequently eased the pangs of separation by sniffing the sweet, musky scent.

Anita repeated her assertion. 'The chemistry *was* right.' Her tone of voice suggested that she was confirming something that she had come to doubt.

Anita and Greg were happy together for about six months. So happy, in fact, that Anita invited Greg to move in with her, which he agreed to do. Anita was a divorcee and she lived with her eight-year-old identical twins.

'How did that work out?' I asked.

'Bradley and Bo love Greg. They got on right from the start. The fact that Greg arrived with an Xbox certainly helped.'

'Do the twins have much contact with your former husband?'

'Not much. They look forward to seeing him but he's very unreliable. He's always letting them down.'

Anita's former husband was a city trader with a cocaine habit. 'I tried to save the marriage but the situation became intolerable.' She saw my look of concern and pre-empted my question. 'No, he didn't hit me or anything like that. Christ, I'd have walked. He just became impossible to live with. Moody – deceitful – I had to think of the boys.'

Soon after Greg moved into Anita's flat their relationship began to deteriorate. 'We stopped talking,' Anita said, her eyebrows tilting severely. 'He doesn't seem that interested in *us* any more. He stays out late and when I text him he never replies.' They were growing apart. 'He's always going out. He's never there for *me*.' Anita became less interested in sex. 'I need to feel close to someone – to be intimate.' And Greg became irritable. 'He called me a control freak.' She gave me a complicit look and laughed. 'I didn't know what to do,' she continued. 'It's a big thing, deciding to live with someone – especially when you've got kids. I thought maybe I'd made a mistake and felt quite low, so I went to see my GP, who put me on Prozac, but I got horrible side-effects and he suggested I come to see you instead.'

Anita was obviously upset but her voice was steady. She

didn't become tearful and knew exactly what she wanted. 'Look, I just want to get things sorted out.'

'Would Greg be willing to see me?' I asked. 'I'd like to see him on his own first. Then we can arrange some joint sessions?'

Anita stood. 'I'll get him to give you a call.'

Couples therapy has murky origins. It was initially developed as part of a Nazi health initiative. Stable, racially pure and large families were considered desirable if the broader objectives of the Third Reich were to be achieved. Needless to say, couples therapy evolved into something quite different after the war. Today, several forms exist, but most share common elements such as training in communication and problem-solving skills. Unhappy couples have few rewarding exchanges but a large number of punishing ones (most of which are angry or accusatory); patterns of reciprocal negative behaviour escalate; sex becomes less frequent; and time spent together ceases to be pleasurable.

I'd noticed that Anita was fond of using words like 'always' and 'never'. Greg was *always* staying out late and he *never* replied to her texts. Use of unqualified language is rarely accurate and usually reflects a style of thinking strongly associated with distorted perceptions. The German-born psychoanalyst Karen Horney was one of the first psychotherapists to connect language with psychological vulnerability. She referred to the 'tyranny of shoulds' to underscore how uncompromising inner speech can create stress and guilt: I *should* be perfect, I *should* be thin, I *should* be successful. Encouraging patients to develop a more nuanced vocabulary can help them to achieve a better match between their inner speech and reality. This simple corrective frequently leads to more measured assessments and improved mood. I made a quick technical note: *overgeneralisation.*

The following week I was able to see Greg – a well-groomed, mild-mannered man with a first-class degree in mathematics from Cambridge. I summarised Anita's grievances and waited for Greg's response. His lips twisted to form a mirthless smile. 'She hasn't told you, has she?'

'I'm sorry?'

He sighed and leaned forward. 'I don't go out much – once or twice a week maybe – and when I do go out I usually text Anita. I tell her where I am and when I'll be back. In the past I did forget to text – that's true – but I wouldn't now. Anita gave me such a hard time. I just don't understand why she's so insecure. She's out of my league really. I mean, she could have been a model.' He looked to me for confirmation and I nodded. 'But she acts as if she doesn't have any options.'

According to Greg, Anita had invited him to move into her flat because she wanted to keep him under surveillance.

'She thinks I'm going to meet another woman and have an affair. But I wouldn't – it's not the way I am. Besides, I love her.'

'Do you tell her that?'

'Sure, all the time, but it doesn't have any effect. She still thinks I'm playing away. She's constantly asking me about where I've been and who I've been with. It's like an interrogation. And if I make a tiny mistake – if she thinks she's found an inconsistency – she gets really upset. And then she shuts down and won't talk.' His chin fell to his chest and his eyes glazed over. 'Gradually, she comes out of it, but I have to reassure her and swear that I'm telling the truth.' He looked uncomfortable and picked at a loose thread on his jacket. 'She demands to see my emails and credit card statements.'

'And you show them to her?'

'I've got nothing to hide. But it isn't right, is it?' He leaned back on the sofa and stroked a neatly cultivated sideburn. 'One night, I got back home and took a shower. I was in the cubicle and Anita came in and removed the laundry basket. She said she was going to run a wash.' His eyes were questioning, uncertain. He wasn't sure whether to continue. 'The thing is – she didn't run a wash. She just wanted to check my clothes.'

'How do you know that?'

'I could be wrong – but I don't think so.'

'She was looking for evidence . . . '

'She's obsessed.'

Greg cringed. Thoughts can be like spells. Their potency is only evident when spoken aloud.

It was necessary to clarify one issue before proceeding. 'Greg, I'm going to ask you a personal question and your answer will be treated confidentially.'

'Okay.'

'Have you been unfaithful?'

'God, no!' He was offended. 'I really want this relationship to work. I've never been unfaithful to anyone. It's not who I am.'

*

Anita fished a rubber band from her shoulder bag, grabbed her hair, and slid it into a ponytail. 'We're different people,' she said. 'We see things differently.' I conceded that sometimes it's difficult to establish objective facts.

'So how often does Greg go out? Is he out all the time – as you suggested – or is it more like once a week?'

'Perhaps I *was* exaggerating. But that's not the point. The point is that we aren't spending enough time together.'

'Have you ever asked to see his credit card statements?'

'Not recently.'

She continued to be evasive but then said, 'All right – I suppose I can be quite controlling, possessive – whatever. But so what? If you love someone, isn't that just natural?'

Jealousy and love are inextricably entangled. The medieval cleric Andreas Capellanus compiled thirty-one rules of courtly love and the second of these is: He who is not jealous cannot love.

'Yes,' I agreed. 'It is natural. Love and jealousy go together. It's a lot easier to be relaxed about fidelity if you aren't in love.'

'Exactly.' Anita appeared relieved. 'It shows you care.' She made a confession. 'I think about infidelity a lot. I have these awful daydreams – they're more like nightmares really. I picture Greg meeting someone in town. They find a seedy hotel and book a room for the afternoon.'

'Who's the woman?'

'I don't know. No one . . . and anyone. I can't get the images out of my head.' She shuddered. 'I even see them in bed together. It's horrible – it makes me feel sick. I feel sick right now just thinking about it.' Her daydreams were accompanied by an overwhelming urge to establish Greg's whereabouts and discover what he was doing. Anita would call his office and if it wasn't possible to locate him she would start to entertain the idea that her upsetting fantasies were real. When I returned to the subject later she said, 'I'm a very intuitive person. Perhaps that's what it is – this feeling – female intuition? I seem to know instantly if I'm going to get on with someone. It's really useful to be able to do that in my line of work. You don't waste time with clients who are going to be hard to please.'

Anita might well have been a good judge of character, but this didn't mean that she had special powers. Nor did it mean that her fantasies had any significance. False inferences inevitably lead to false conclusions.

I was still unclear about what Anita thought was really happening when she couldn't locate Greg.

'Do you believe that Greg is having an affair?' I asked.

'He could be,' she responded.

'That's the answer to another question.'

She crossed her legs and seemed distracted by the long, pointed heel of her boot. She reached out and touched it – almost a caress. 'Sometimes I do and sometimes I don't.'

Jealous protagonists appear in all literary traditions. Euripides' Medea poisons her rival and slaughters her own children; Shakespeare's Othello smothers Desdemona; and Tolstoy's Pozdnyshev despatches his wife with a dagger thrust beneath her ribs. These depictions reflect a grim reality. Estimates vary according to time and place, but broadly speaking the murder of partners and former partners accounts for approximately one in ten of all murders worldwide. The motivation for such murders is predominantly proven or suspected infidelity. Men are more likely to kill women, but women also kill men – albeit in very much smaller numbers. About a third of women who are murdered around the globe are killed by their husbands or boyfriends – often stabbed or beaten to death. A woman is statistically much safer getting into bed with a total stranger than with someone she knows. Although jealousy arises along a spectrum of severity, even mild forms can be explosive.

Throughout the course of the twentieth century, Anita might have attracted any of the following diagnostic labels: Othello syndrome, erotic jealousy syndrome, morbid jealousy, psychotic jealousy, paranoid jealousy, obsessional jealousy and delusional jealousy. Today, these terms have been superseded by Delusional Disorder: Jealous Type – which makes pathological jealousy a close relation of de Clérambault's syndrome (now designated Delusional Disorder: Erotomanic Type).

Whereas the defining feature of de Clérambault's syndrome is a delusion of love, the defining feature of Delusional Disorder: Jealous Type is a delusion of infidelity. The same drugs are prescribed for both conditions, suggesting common neurochemical pathways. There may also be further overlap insofar as pathological jealousy, like de Clérambault's syndrome, has been linked with damage located on the right side of the brain. The two conditions have behavioural similarities. Both types of patient will stalk, although for quite different reasons. A de Clérambault patient does so because he or she cannot tolerate separation, whereas the pathologically jealous do so in order to spy; both have strong intuitions and treatment outcomes are generally poor.

I mentioned to Anita that some of my patients found it easier to control their jealous thoughts when given medication. She was extremely resistant to the idea because she had suffered bad side-effects after taking Prozac; however, there was also another reason. 'Taking pills made me feel different, a bit dead inside. It was really weird, I don't know if I'm imagining this, but they seemed to interfere with my ability to work. I'd walk into a room and nothing would come into my head – you know, colour schemes, textures, materials ... Usually, I get lots of ideas.' This was a complaint I'd heard many times before from patients whose professions required the exercise of an artistic gift. There is a school of thought that mood disturbance, particularly oscillating mood disturbance, promotes creativity. The lifetime prevalence of mood disorders in writers, for example, is significantly greater than in non-writers matched for age, sex and education. Mood swings might plausibly underpin optimally productive cycles of reflective melancholy followed by enabling surges of energy. The artificial stabilisation of mood by chemical means would negate this effect.

Anita had had relatively few sexual partners. She was – in her words – 'quite fussy'. On every occasion when she had permitted lovers to get close, she had been overwhelmed by a sense of her own vulnerability. When she talked of feeling exposed, I noticed that her voice lost strength, particularly at the end of sentences – trailing off into breathlessness. It happens when people are frightened or anxious. The effect was to make her sound like a child.

The idea that we all carry within us a sub-personality, a remnant of who we once were, has its origins in the analytical psychology of Carl Gustav Jung. In 1934 he wrote: 'For in every adult there lurks a child – an eternal child, something that is always becoming, is never completed, and calls for unceasing care, attention, and education.' Jung's precept was subsequently borrowed and reworked by numerous practitioners, particularly during the 1960s and 1970s, and it has since been enthusiastically endorsed by pop-psychologists who frequently exhort us to love our 'inner child'. Overuse and sentimentality have undoubtedly devalued the concept; nevertheless, it is a serviceable way of thinking about how the mind works. Emotions or situations that remind us of significant moments in infancy activate dormant memories and we feel like a child again.

I fancied that I'd heard the voice of Anita's inner child: a little girl, hiding behind drawn curtains, peeking through the gap. I could have encouraged her to step out, but I was apprehensive. She might get scared, run away and never come back again.

You can tell a great deal about a couple by observing where and how they sit. All too often, I saw couples who, on entering my consulting room, would choose to sit at opposite ends of the sofa. It was as though they were surrounded by mutually repelling force fields. When seated, they would lean away from each other or sit at oblique angles so that they were almost back to back. Sometimes, I would ask them to sit closer and they would oblige. But by the end of the session they had usually edged apart again. All of them had expressed a wish to save their relationships, but their words were never quite as informative as their body language.

Anita and Greg appeared to be comfortable with closeness. They sat side by side, sometimes touching.

The role of a psychotherapist conducting couples therapy has much in common with that of a referee or adjudicator. It's surprising how difficult it is for couples to agree on even the basic things – such as who said what and when. The same event is remembered quite differently. They make inferences on the basis of little evidence and are frequently wrong; they act as if they can read each other's minds and reach erroneous conclusions that are then treated as indisputable facts. This can be very frustrating. In therapy sessions, couples become irritable and talk over each other. Highly dysfunctional couples will start trading insults. On many occasions I've been forced to raise my voice and bark commands to get them to stop.

Anita and Greg were bickering, rather than arguing.

'I don't know what more I can do,' said Greg. 'I let you know where I am and I tell you who I'm with.'

'Do you?' Anita asked.

'Yes.'

'That's just not right, is it? You came home late on Tuesday. And you *didn't* text.'

'Oh come on, Anita. Please. That was because of exceptional circumstances.' Greg looked at me and made an impatient gesture. 'There was a crisis at work. I had to attend an emergency meeting. I didn't have the time.'

'How long does it take to text?' Anita cut in.

'I just couldn't,' Greg replied.

'You promised.'

'Anita,' I said, raising my arm to get her attention. 'What *did* Greg promise, exactly?'

'That he'd text – always.'

'That's true,' Greg admitted. 'I did say that, but what you don't seem to appreciate, Anita, is that sometimes – with the best will in the world . . .' His plea became an exhalation articulated by lips that opened and closed without producing any more words. He had lost the will to continue mid-sentence.

I addressed Anita again: 'Do you doubt that there was a crisis?'

'She knows what happened,' Greg said. 'I've shown her my e-mails – and I'll show you too if you like. It was a serious situation and I had to deal with it there and then.'

'Anita?' I prompted.

'It would have taken him a few seconds.'

'People sometimes forget things when they're stressed . . .'

'Not important things.'

'Is that what Greg's oversight means? That you're not important to him?'

'That's what it feels like.'

'Which must be very upsetting. But I'm asking you now for a *considered* answer. Do you *really* believe that because Greg didn't text you, on this particular occasion, that you're not important to him?'

I was trying to get her to stop and reflect, to acknowledge the possibility that some of her thoughts might have become automatic, examples of what cognitive therapists sometimes call thoughtless thought; however, I'd asked the question too emphatically, the stresses were wrong, and I'd made her feel as if she were being cross-examined like a witness in a court of law. I had also previously referred to Greg's failure to text her as an 'oversight', which suggested, perhaps, that I had already decided that his behaviour was excusable. A psychotherapist, like a writer, must use words with great care.

Anita tensed. 'Look,' a flush of anger reddened her cheeks. 'Greg's always telling me that I can rely on him. That he's trustworthy. But I can't even trust him to send a text.'

Couples frequently argue over things that appear entirely unimportant to others, going round and round in circles without resolving anything or getting anywhere. They are like medieval theologians, interminably debating the number of angels that can dance on the head of a pin. But when couples clash over seemingly minor issues it's worth straining to hear the subtext. It's not the argument that's important, it's what the argument reveals.

Greg was also getting annoyed. 'This is ridiculous,' he grumbled. 'Anita, you're getting everything completely out of proportion. I got home – what? – an hour late?'

'An hour and ten minutes.'

'Fine.' Greg rolled his eyes. 'An hour and ten minutes.'

I added a single word under my existing scrawl of notes: 'perfectionism'. And then I underlined it.

Why do people get jealous? If you love someone you should want them to be free and happy. True love knows no bounds; it releases the soul – takes us beyond conventional limitations. The Lebanese poet Khalil Gibran, whose works are among those most frequently read at wedding ceremonies, wrote: 'Love is the only freedom in the world because it so elevates the spirit that the laws of humanity and the phenomena of nature do not alter its course.' These are uplifting words, but Lucretius gets much closer to the truth when he warns us that the goddess of love has sturdy fetters. We are only free to be ourselves and that isn't very free at all.

Utopian communities have adopted 'free love' as a guiding principle, but virtually all of them have dwindled or collapsed on account of group members reverting to monogamy. Where polygamy is permitted, only 5 to 10 per cent of men choose to have several wives. The internet has opened up a channel of communication between young couples eager to explore a 'polyamorous' lifestyle, yet many of them confess that overcoming jealousy is a major obstacle. Couples who manage to maintain stable 'open' relationships and raise children constitute only a tiny fraction of the general population. Whenever social engineers or political visionaries have attempted to alter the structure of society, the family unit returns. Our need to privilege a single, exclusive relationship and guard it jealously is clearly hardwired.

In the ancestral environment, a thriving infant returned a healthy premium on parental investment: the survival of their genes. For mothers, the most significant threat to achieving this end was the dispersal of family resources – something likely to happen if her partner mated with another female. For fathers, the cost of a partner's infidelity was far greater. He might squander all his resources furthering the prospects of another man's genetic material. Jealousy is an alarm that triggers preventative manoeuvres – radar for rivals. Given that the cost of infidelity is so much greater for men, male sexual jealousy is generally more intense, which explains the marked gender asymmetry observed with respect to spousal homicide statistics.

Anita's alarm was going off all the time.

'When Greg was still living in his old house, I used to let myself in early so I'd have time to check.'

'You went through his things?'

'No . . .' I waited and she returned my gaze. Horizontal lines appeared on her brow and she touched her chest as if troubled by a palpitation. 'I used to check his bed.'

'What were you looking for?'

'You know . . . stains, hairs.'

'Traces . . .'

'Yes.'

'And did you find any?'

'There are always hairs in beds. I used to pick them off the sheet and hold them under a lamp . . .'

'Were they the kind you were looking for?'

'There's always some doubt.'

'What else did you do?'

'I'd sniff the pillow, for perfume.'

She had been looking for a woman who wasn't there. And no matter how many times Greg assured Anita of his innocence, she continued to look for evidence of her elusive rival with the thoroughness of a forensic scientist.

In some instances of extreme jealousy, the affected individual will continue asking questions, monitoring and checking, even after a spouse has actually confessed to being unfaithful. This suggests the existence of a neurological 'switch' that has become stuck. Under such circumstances, jealousy resembles obsessive-compulsive disorder (OCD), a condition in which intrusive thoughts produce a state of anxiety and discomfort that the individual attempts to reduce by performing rituals. These rituals are compulsive and associated with strong urges that are hard to resist.

An effective treatment for OCD is exposure and response prevention. The individual is asked to tolerate discomfort in situations which create anxiety, while resisting the urge to perform ritualistic actions. Scanning studies have shown that this form of 'behaviour therapy' reduces activity in certain areas of the brain such as the caudate nucleus and the thalamus. These altered patterns of brain activity are almost identical to those observed when OCD patients take medication. Remarkably, biological abnormalities associated with psychopathology can be corrected through the exercise of willpower.

The majority of OCD patients treated with exposure and response prevention subsequently experience fewer intrusive thoughts, become less distressed and feel less compelled to perform rituals.

I asked Anita to resist performing her checking behaviours. Although she could resist for short periods of time she was simply unable to sustain her efforts. A disturbing image of Greg and another woman would come into her mind and her malignant suspicions would metastasise until she was consumed with jealousy and overcome by compulsive urges to resume her questions and detective work.

*

Almost all schools of psychotherapy agree that stresses experienced during childhood have long-term consequences for mental health. Some carry this principle to its logical extreme and assert that even pre-natal experiences are of critical importance. Foetuses evidently experience and react in the womb, sometimes prefiguring adult behaviour. Ultrasound scans show that after only twenty-seven weeks of gestation, baby boys' penises become erect when they suck their thumbs.

I asked Anita about her childhood.

'Mum's an artist,' she said, without pride. 'Big, colourful abstracts . . . she's always been very dedicated, even though she's never sold enough work to make a proper living. She didn't have much time for us when me and my brother were kids – fancied herself as a Bohemian – still does. Men would come and go. I was only small but I knew something was going on; I knew something wasn't right. My mum used to make me and my brother accomplices.' Anita's eyes expanded and she spoke like a medium channelling an inebriated ghost: 'Don't tell your father.'

'Did you keep her secret?' I asked.

'Yes,' Anita replied. 'Of course – but Dad found out somehow. She might even have told him herself. It was the sort of thing she would have done in the middle of an argument – for effect – she loved drama. They were always having rows, breaking up and getting back together again. My brother and I got used to being packed off to our grandparents when they were patching things up.' Anita's eyes misted with recollections. I sensed a curious duality, a vague impression that the adult Anita and the child Anita were suspended in a state of indeterminacy – like a quantum superposition flirting with prospective realities. 'Eventually, Mum settled down and the men stopped coming. I can't say why – maybe she just got too old for it all. They seem to get on well enough now, Mum and Dad. But it was difficult – back then.'

Somewhere in the building a phone started purring.

'How do you feel about your mother?'

The purring stopped.

'It's difficult. She's my mum – but she was a pretty useless parent. She just wasn't interested. I think she found looking after children dull.'

Donald Winnicott, a luminary of British postwar psychoanalysis, asserted that the mother-child relationship is the principal safeguard against future mental illness. Indeed, he went on to argue that ultimately the selflessness of ordinary mothers makes the whole of civilisation possible. A central feature of Winnicott's theory of emotional development is 'holding' – a term that he employed to mean not only physical holding, but every aspect of maternal care: feeding, bathing, attending, comforting. A 'holding environment' makes the infant feel safe and facilitates the transition from dependence to independence. Holding is also the first experience of relating and it informs all subsequent social activity.

Psychotherapy is a kind of holding, a safe place in which to explore and grow. The relationship between a therapist and patient, when it works, replicates good parenting. It is a vehicle for encouraging transitions.

After seven weeks, Anita had acclimatised to psychotherapy. She was disclosing with less apprehension and I judged that she could cope with more probing questions. I wanted to excavate her jealousy; to plumb its depths. A useful technique for achieving this goal is what cognitive therapists call 'downward arrow'. It is usually represented in diagrams as a vertical line of downward-pointing arrows separated by questions that vary slightly but are actually the same, insofar as they are all inquiries about meaning.

'Anita, have you ever thought about how you'd react if Greg admitted to having an affair?'

'I'd be devastated. You don't think that, do you? He hasn't said anything, has he?'

'No, not at all,' I reassured her. 'I'm just interested in what that would mean to you – if it were true.'

'Mean?' She looked puzzled for a moment and then said: 'It would mean that he'd been lying all along.' She shifted and looked at me with suspicion. I was asking her to state the obvious.

'And what would that mean?' I continued.

'He can't be trusted.'

'And if that were true?'

'Christ – if you can't trust your partner – well, who can you trust?'

'Okay. Let's say you can't trust Greg – or anyone else you might meet in future. What would that mean?'

'That there's no such thing as closeness.'

'And if that's true?'

Her words were given intermittent sustenance by a slow, tremulous breath. 'I'm alone.'

She looked very frightened. The inner child had finally stepped out from her place of concealment; a child who had looked to a selfish mother for love – a mother who was as *unfaithful* to her children as she was to her husband – only to be ignored and rejected.

Evolution has ensured that we form a close bond with our parents, because on the plains of Africa, an abandoned child was a dead child. Anita's rival – the woman who wasn't there – was more than a sexy temptress. She was Death's understudy. The prospect of betrayal terrified Anita, because betrayal deposited her inner child in the middle of an ancestral wilderness of lengthening shadows and predatory darkness.

Greg was sitting a short distance from Anita.

'There's no need to ask me about my past. It's done – gone – finished. It's the past.'

'Why must you always be so cagey about it?' Anita responded.

'How can you say I'm cagey – I mean, really . . . '

'I've told *you* everything.' The heavily weighted pronoun made the sentence an accusation.

'I know you have. But there was no need. Why should I care about how many lovers you've had?'

'Openness – honesty? These things matter.'

'Oh, come on . . . this isn't about honesty and openness.'

'Then what is it about?'

'When you ask those questions . . . I feel like I'm being manipulated.' The atmosphere chilled. Greg looked to me for assistance, but I simply revolved a finger in the air, encouraging him to continue. 'It's a pretext,' he added.

Anita's response was sharp: 'What?' Almost a shriek.

'It's a pretext. You say let's be open – but it's not really about that. It's about getting more information – so you can compare and contrast – so you can trip me up. And the thing is, you're always going to trip me up, because I can't remember *everything* – not exactly. There are always going to be mismatches. It doesn't mean that I'm lying. It doesn't mean that I'm trying to mislead you. It just means that I can't remember, because these things – my past relationships – they aren't important to me any more. You are!'

The corners of Anita's mouth had turned down slightly.

'What are you thinking?' I asked her.

'Is it too much to ask?' Anita's question was addressed to the ceiling rose. 'All right – I have a problem.' She turned to Greg. 'But maybe if you tried a bit harder . . .'

'Me?' Greg thumped his sternum. The impact made a loud thud. 'Me – try harder? I'm not sure that's possible. Besides, whatever I do will never be enough, whatever I say – whatever I do – you'll never be satisfied.'

There is nothing wrong with being a perfectionist, but if internal standards are too high this otherwise commendable trait can interfere markedly with general functioning. It is associated with several psychiatric conditions, most notably anorexia nervosa, OCD and depression. Opinions differ widely concerning the nature of perfectionism. At one end of the spectrum are the views of psychoanalysts, who understand perfectionism as a defence against harsh parental criticism, and at the other end are those of cognitive scientists, who understand it as a set of largely unmotivated preferences that arise directly from the brain (like an instinctive need to line things up, for example).

Anita's mother was not only neglectful, but highly critical. In a previous session, Anita had said: 'Mum was constantly finding fault. I might have become an artist – like her – if it wasn't for her remarks. She was always being negative.'

I was inclined to view Anita's perfectionism as motivated rather than unmotivated; a defence, rather than a property of her nervous system – although these contrasting approaches are by no means mutually exclusive. Sometimes, a single characteristic can be overdetermined, the result of several causes.

Anita would continue ruminating about Greg's past and asking questions, because he could not supply her with *perfect* answers. As someone conversant with coding, Greg would have recognised that Anita was caught in a 'test-operate loop' with no 'exit'.

Anita was looking miserable.

Greg had more to say: 'Even when I'm just sitting quietly you act like I'm doing something wrong.' He turned to face me again. 'At home, sometimes I'll be a bit preoccupied – you know – just thinking about stuff, and Anita will get agitated and say: "What's the matter? Why aren't we talking?"' He took Anita's hand and moved his thumb backwards and forwards over her knuckles. She raised her head and looked into Greg's eyes. 'I want us to be happy together,' he said. 'But what you do – it's oppressive.'

The taproot of Anita's jealousy was a child's terror of abandonment, a primal fear selected by evolutionary mechanisms and amplified by early learning experiences. Ultimately, it was this fear of being abandoned that made her vulnerable to developing a psychiatric illness. But what does this mean, to be psychologically vulnerable? It's relatively easy to think of physical vulnerabilities as weaknesses attributable to vitamin deficiency or brittle bones. But what is the psychological equivalent? What do the causes of psychological vulnerability look like?

Cognitive psychologists refer to 'schemas' or 'schemata' when describing connected beliefs that influence the way we see, understand and react to the world. They are created by learning experiences.

The idea of a schema might be quite difficult to grasp because it is a hypothetical construct and cannot be observed directly; however, one might think of a dysfunctional schema as a metaphorical lens through which the incoming light of experience is refracted by flaws in the glass. The subsequent distorted image is inaccurate and arouses strong emotions. For example, if the distortions make the world look as though it is full of monsters and dangerous, we will experience irrational fear.

Aaron T. Beck – the American psychiatrist who developed cognitive therapy – has suggested that dysfunctional schemas store knowledge in at least two ways: first, as conditional propositions or assumptions, e.g. 'If I am not loved then I can never be happy'; and second, as unconditional statements such as 'I am unlovable'. The latter is an example of a 'core belief'. Core beliefs are more defining than assumptions and are buried deeper in the psyche.

Many important learning experiences precede language acquisition, so some schemas are either entirely non-verbal or they incorporate non-verbal elements. In the absence of language, learning is embodied. When our bodies 'remember', we experience physical symptoms, such as an accelerated heart-rate, hyperventilation, or butterflies. Perhaps this is why we frequently say things like 'I can feel it in my bones' or 'I have a gut feeling'. Elusive perceptions seem to register not in the brain but at any number of other sites around the body.

Schemas exert their influences from below the awareness threshold. When activated, pre-verbal schemas will reproduce the powerful, raw emotions of infancy. A trigger, such as jealousy, might activate an 'abandonment schema', and the vulnerable individual will then be overwhelmed by a sense of terrible isolation. This happens automatically with very little cognitive mediation. The principal task of psychotherapy is to make the vulnerable individual aware of his or her schemas and to modify them by correcting dysfunctional assumptions and harmful core beliefs. The therapeutic relationship – which frequently requires the therapist to act as a surrogate parent – can be an important catalyst for change at the pre-verbal level; however, such change is extremely hard to achieve and usually requires a long-term commitment from both therapist and patient.

An abandonment schema is activated almost exclusively during the course of an intimate relationship; a partner's neutral remarks and behaviour are misinterpreted in a negative way and the affected individual overreacts. A panicky, highly aroused state is followed by a state of detachment and withdrawal, which can also serve as a means of punishing the unwitting partner.

Schema-focused cognitive therapy is an extremely good example of convergence among therapies. Although cognitive therapists and psychoanalysts have very different practice methods, the principal objective of both approaches is to reduce the power of influences emanating from the unconscious and increase awareness of the origins of self-defeating behaviour. The individual will then be able to make more accurate appraisals that will inevitably inform rational, reality-based judgements.

I was feeling relatively optimistic. Although treating jealousy can be challenging, we were making progress. Greg and Anita were communicating and they were both very motivated and emphatic about wanting their relationship to succeed. I had a formulation – a rather satisfying diagram in my notes – showing the proximal and distal causes of Anita's jealousy, with arrows connecting core beliefs, assumptions and thoughts, circles showing how certain behaviours were maintained, and little boxes detailing modulating variables. Anita was a textbook case, conforming to the expectations of various psychological models and theories. It followed that therapy should be effective.

The next day Greg and Anita broke up.

I saw Greg first. A small plaster was stuck to his forehead, just over his left eyebrow.

'On Friday night we met with some friends at my tennis club. Anita was relaxed and we were having a great time. We were messing about, laughing and joking. Anita can be really good in social situations. You'd never guess. You only see her when she's in here – talking about our problems. But out there,' he glanced towards the window, 'she can be a lot of fun. There was a woman standing at the bar. She looked familiar and when she turned round I recognised her – it was Kate, an ex-girlfriend of mine. We were together four, five years ago – and not serious. I kept my head down hoping that if I ignored her she'd just walk past, but Richard, one of the guys we were with, called out her name and invited her to join us – she works for the same travel company. Anyway, I didn't quite know how to handle the situation. Kate was friendly and it was obvious we knew each other, but no one was asking *how* we knew each other – and I think Kate sensed that things were a bit awkward because we didn't talk for very long. She spent most of the time chatting to Richard and his wife. Thankfully, Kate left as soon as she'd finished her drink. From that point onwards Anita didn't say very much. She was very quiet for the rest of the evening. I was expecting trouble and probably drank too much.' He made an appeal for clemency with guilty eyes. I responded with an absolution sketched in the air with priestly benevolence and he continued: 'I'd had a busy week and I wanted to have a quiet, uneventful weekend with Anita and the boys. I didn't want to argue. Going back home in the car, things were quite tense. Anita said: "How do you know Kate?" I told her that Kate was an ex and then she said: "So when were you going to tell me?" And I didn't have an answer to that, because what I really wanted was to just forget it and go home and make love – like normal people. Anyway, I explained that it had been a very brief affair and I hadn't seen Kate for years and Anita said: "What? You've never seen her before at the club?" And I said: "No, I haven't." But Anita was suspicious and I could see that she was getting more and more upset.' Greg looked at the empty space next to him on the sofa, as if he had suddenly become aware of Anita's absence.

'We got home – I paid the baby sitter – and while we were still in the kitchen Anita started asking more questions, one after another – questions, questions, questions. It was relentless. And then she said something utterly ridiculous like: "Do you still find her attractive?" And I said: "Yes, I still find Kate attractive." I was going to add "but I don't love her" when Anita picked up a plate and threw it at me. She missed but it smashed against the wall and a piece caught me here.' He touched the plaster on his forehead. 'And I thought to myself – I just can't go on. I just don't need this. This is no way to live.'

'What did Anita do?'

'She started crying and then Brad woke up and Anita had to get him settled again. When she came back I told her that I thought it would be best for all of us if I moved out.'

'How did she react?'

'She just went into herself. She became – I don't know – blank – frozen. ' Greg brushed away a tear.

'It's okay . . .' I said, placing the tissue box on the sofa.

He looked at it for a few moments and then snatched a tissue. 'She's so beautiful. I mean – amazing.' He blew his nose and stuffed the tissue into his pocket. 'Whatever we had is buried under so much shit now. I hope the boys are going to be okay. They're great kids and I'll miss them. But if Anita and I stay together and she gets as mad as she did on Friday night – that isn't going to be good for any of us. Brad and Bo, they shouldn't see their mother behaving like that – they just shouldn't.'

We discussed options: a temporary separation, more intensive therapy. But Greg was sure. The relationship was over.

'You might change your mind,' I suggested.

'No,' he said firmly. 'I want my life back.'

I saw Anita the next day. She entered the consulting room with an air of brisk efficiency and business-like purpose. Her hair was pulled back with an Alice band and she was wearing more make-up than usual. It had been applied skilfully, but the effect was of unnatural smoothness – artificiality, the complexion of a mannequin or a doll. She sat down, crossed her legs and recounted her version of events.

'I was suspicious the moment I saw her. There was something . . . something going on between them. I could just tell.'

Greg had been shifty – he'd refused to answer her questions – and yes, she had lost her temper – but his behaviour had been unforgivable.

'If I hadn't asked him about her, he wouldn't have said a word.' The whole episode had been utterly humiliating. She had been obliged to sit at the same table as Kate, 'Watching her catch his eye – watching her flick her hair.' Anita did an extraordinarily good impression of a manipulative woman capturing a man's attention while feigning innocence. 'How could he have done that to me?'

There was a slender possibility that Greg had decided against ending their relationship and I thought it wise to remain silent on the matter – at least until I knew what had actually transpired.

'You saw him yesterday – didn't you?'

I felt a little duplicitous. 'Yes. I did.'

She set her jaw defiantly before saying: 'We're breaking up.'

'I see.'

Somewhat breathlessly, Anita added: 'He's leaving me.'

She didn't make a sound at first, but slowly, sobs came, and then a great, heaving cry of anguish, which made her snap forward as though she'd been winded by a hard punch. Tears coursed down her face leaving vertical trails of mascara. With the unguarded, careless urgency of a child, she slid the back of her hand under her nose to wipe away the snot. I tried to engage her, but she had regressed to some pre-verbal state where horrors are necessarily unspeakable, where words cannot be interposed between a chaotic world and the self, where language isn't accessible to give despair a name and thereby contain it.

When I was an undergraduate, I was discussing my ambition to train as a clinical psychologist with one of my lecturers. 'So,' he said, 'you fancy the misery game, do you?' He was being purposely provocative. 'A life in misery . . .' His intention was to make me give proper consideration to the potential effects of spending the rest of my life in enclosed spaces observing people suffering. Spectacular grief can be cloying – hard to shake off.

The only time I have ever come dangerously close to crying in a therapy session was listening to a boy recounting the circumstances of his mother's death. What had started off as an exciting day out for his family had ended in disaster. They had found themselves caught up in a series of events, the outcome of which was a national tragedy. Hundreds were injured and many people were killed. It wasn't pity aroused by the boy's distress as he described screams and disfigured corpses that moved me, but rather his courage, his dignified efforts to remain composed because he wanted to give an accurate account of his mum. He wanted to affirm her character and kindness, to give her short life meaning by telling others about her.

Anita's sobbing subsided and she became very still. The adult surfaced and she said, 'I didn't mean to hurt Greg. I really didn't. I just went into a blind rage.'

Why do people engage in self-defeating behaviour patterns?

Freud employed the term repetition-compulsion to describe an innate tendency to reproduce early traumas in the context of current relationships. The outcome that Anita most wanted to avoid was abandonment, the condition of her childhood, yet she persisted in behaving in ways that made Greg's departure an increasingly likely outcome. Although much of her behaviour was motivated unconsciously, she must have had some awareness of what might happen. The division between the unconscious and conscious is not absolute; there are twilight regions, penumbral fringes and blurred boundaries. Moreover, she was perfectly aware of her history. She had had relationships with other men before Greg, she had experienced jealousy, made accusations, and subsequently, these relationships had also come to a premature end. Why did she carry on repeating the same mistakes? Why was she so inflexible?

Consideration of the origins of repetition-compulsion eventually led Freud to infer the existence of the death instinct, a drive that lends support to all forms of self-defeating behaviour and ultimately self-destruction. He justified the notion with recourse to a law of nature: organisms evolve from inanimate matter and must inevitably return to the inanimate state. This common destiny finds correspondences in our thoughts and predispositions. When we engage in self-defeating behaviours, we are allowing the death instinct to carry us a little closer to oblivion.

Repetition-compulsion is probably more economically explained as a kind of bad habit. We learn certain patterns of behaviour very early and they become our default setting. These behaviours arise from schemas that are so entrenched, so central to our sense of self, that any departures from their script make us feel wholly disorientated. We experience what the radical psychiatrist R. D. Laing called ontological insecurity; we no longer perceive the world as a place of unquestionable, self-validating certainties. We feel like we're losing ourselves.

Self-defeating behaviours persist – even when they cause us pain – because the alternatives are associated, at least initially, with greater distress. Dysfunctional schemas are like an old pair of shoes. They aren't really fit for purpose, but they're what we're used to and they don't pinch.

When a relationship breaks down, couples can still benefit from attending joint therapy sessions, particularly if children are involved. There are usually loose ends to tie up, bills to be paid, outstanding issues to be resolved before all parties – including the children – can move on. Civil communication is still necessary if couples are to separate without causing too much collateral damage. I spoke to Anita about this possibility. She wasn't interested and nor was Greg.

I suggested to Anita that she might consider making a longer-term commitment to psychotherapy. She said she'd think about it, but I wasn't convinced.

'Perhaps you're feeling let down right now.'

'You tried . . .'

'It would be understandable.'

'I'm disappointed. But I don't feel let down.'

We worked on her depression and talked a great deal about trust.

'If I'd known for certain,' she said, 'that Greg was trustworthy – then I wouldn't have had to ask him so many questions.'

'But how can you ever be certain? There are no guarantees. When we love we have to take risks.'

'I can't take risks.'

'Other people do.'

'I'm not other people.'

She studied her pointed heel and tested its point with her finger.

Anita wanted love, but to love is to be jealous, and to be jealous in the way that Anita was precluded love. After she broke up with Greg, Anita came to see me for six more sessions and then cancelled the following three. The final page of her notes: three dates followed by the letters D, N and A.

I never saw her again. Or at least, I never saw her again in person.

The *Diagnostic and Statistical Manual of Mental Disorders* (DSM) of the American Psychiatric Association is a comprehensive guide to diagnosis and classification. It is currently in its fifth edition (DSM-V). This diagnostic bible has attracted a great deal of criticism over the years, the most serious of which are that it is non-empirical and that the content is overly influenced by drug companies. Many psychologists and psychotherapists believe that the entire enterprise of psychiatric diagnosis is misconceived, simplistic, misleading, reductive and prejudicial. People – they say – should not be 'labelled'.

Personally, I have no ideological objection to diagnosis. A diagnosis is nothing more than a summary term for symptoms that tend to cluster together. Some diagnoses are less convincing than others, and there is always a danger of medicalising normal behaviour, but on the whole, I find diagnoses useful, and classification a means of imposing order on what would otherwise be a bewildering and confusing universe of symptoms. I prefer DSM to its competitor, the ICD system of the World Health Organization, simply because I find it easier to read and more digestible. There is, however, enormous overlap between the two.

As mentioned earlier, Anita easily met DSM-V criteria for a diagnosis of Delusional Disorder: Jealous Type. This condition is classified under the section heading 'Schizophrenia and Other Psychotic Disorders' – a group of very serious mental illnesses. But how can one ever be sure that suspicions are delusional? Without constant monitoring, twenty-four hours a day, it is impossible to be certain.

Some 20 to 40 per cent of married heterosexual men

admit to having had at least one extra-marital affair – as do 20 to 25 per cent of heterosexual married women. Approximately 70 per cent of dating couples cheat on each other. Over half of the single population engage in 'mate poaching' – attempting to break up an existing committed relationship. From the perspective of evolutionary psychology, the human reproductive strategy is mixed, a judicious combination of pair bonding and opportunistic sex.

The diagnosis of Delusional Disorder: Jealous Type is actually very speculative. It depends entirely on a decision about the unknowable. Was Greg telling the truth? Had he always been faithful? I thought he was an honest, decent man. But I could be wrong. He might have been a skilled manipulator with obscure motivations. And if he was lying, where does that leave the diagnosis? It is a worrying thought. I reassure myself by going over the facts, by considering what I *do* know, rather than ruminating about what I don't know and can never know. Anita had deep-rooted problems and a history of pathological jealousy. She had no substantial evidence to support her unhappy conclusions. As such, she could be described as delusional. At the back of my mind, however, there is a persistent, niggling doubt.

Some ten years after I watched Anita leave my consulting room for the last time, I was lying on a bed in a hotel room in New York, pointing a remote controller at a wall screen and channel hopping in a distracted fashion. I glimpsed a face that looked familiar. It was definitely Anita. She hadn't changed much at all. Her eyes were unmistakable. She was standing in the middle of a large, opulent room talking about colours and fabrics, and I realised I was watching a trailer for an interior design show. I jumped off the bed and tried to see if Anita was wearing a wedding ring, but the image dissolved and I found myself studying a map of the east coast of the USA and listening to a weather forecast.

Chapter 4

The Man Who Had Everything: *Addicted to love*

In therapy, some patients engage in what might be likened to an emotional striptease. Layer by layer, resistances are removed, until the final concealments fall away and the truth, however painful, unpalatable, or shocking, is revealed. That moment, the moment before the final revelation, is hyper-intense.

Almost thirty years ago I was seeing an entrepreneur – a lean, elderly gentleman with a Van Dyke beard and a fondness for colourful waistcoats – for stress management. He told me about a project that he was about to invest in but I didn't really understand what he was talking about. A few years later I was able to decode what he'd been saying. It was a project that effectively changed the world. He came to see me four times.

The first three sessions were routine – an assessment, a formulation and some preliminary educational work.

He was an affable man with working-class origins and like many people who have ascended the social hierarchy to positions of influence and power he was fond of telling stories that highlighted the magnitude of his achievements. I had to keep reminding him that we had a job to do. He had a heart condition and his cardiologist considered stress management an important component of his long-term care. He would smile and make bountiful gestures: *What's the hurry? We have all the time in the world.*

A false smile engages the muscles around the mouth but not the muscles around the eyes. The delta of lines that fanned out across the entrepreneur's temples didn't move. In fact, they never moved.

When he arrived for his fourth session, his manner was more subdued. His answers to my questions were elliptical and eventually he reached for a tissue. The sagging pouches beneath his eyes had collected a few tears. I asked him what was wrong and he continued to give me vague, unsatisfactory answers until the session was almost over. He glanced up at the wall clock and then studied me with such concentrated attention that his brow became compressed and striated. We had less than five minutes. 'Stress management, eh?' He said the words with a hint of detraction. 'Some forms of stress can't be managed.' His pale grey eyes, slightly cloudy in appearance, didn't blink. I could hear the rush of blood in my ears. The moment was suspenseful, like the charged latency that precedes a thunder clap. 'A long time ago,' he continued, 'I was on a boat in the middle of the Arctic Ocean – skirting the pack ice – and I ordered a man to be thrown overboard.'

'Who was he?' I asked.

The entrepreneur replied with solemn determination: 'He was a very, very bad man.'

'And you left him there?'

'Yes. As I said, he was a very bad man. Do you understand? Very bad.'

Was my patient being serious? Or was this some kind of test? Perhaps I was being duped?

'We're going to have to talk about this . . . '

The entrepreneur tugged at his cuffs and showed me the time on his wristwatch. The hour was up. 'I have another appointment.' He stood, adjusted the hang of his trousers, put on his long coat, and shook my hand. 'Yes,' he said. 'We'll talk about it.' After he'd left my consulting room I looked out of the window. A black Mercedes was parked outside the entrance on triple yellow lines; a uniformed chauffeur got out, opened one of the rear doors, and I watched as the entrepreneur vanished into the shadowy interior. He didn't come back.

Psychotherapists sometimes refer to 'admission tickets' – relatively trivial problems that patients use as a pretext for entering therapy. When the patient feels comfortable and emotionally safe, the actual or substantive problem is revealed. It isn't uncommon to discover that the real problem has some moral dimension which is the cause of a troubled conscience. There are marked parallels between psychotherapy and the Catholic confessional. Secrets weigh heavily on the soul and the process of unburdening can be a great relief. For some people, psychotherapy is all about confessing.

*

Ali was a man in his late thirties. His grandfather – who had arrived in the UK as a penniless immigrant – and his father (who had died when Ali was young) were responsible for creating a large and profitable manufacturing business. Ali was a family man, a good man, respected by the leading members of his community. I found him friendly enough, but he had about him a certain air of detachment. It was something I was familiar with – this manner, this *attitude* – which made me feel that my patient was seated behind a thick sheet of glass. Ali was slightly removed and emotionally flat. I didn't feel that we were connecting properly.

For treatment to be effective there needs to be some kind of connection between therapist and patient. It can be a simple bonding, as might develop in any situation where two people work together to achieve the same ends. Or it might be a more complex connection, as in psychoanalysis, which gives special significance to the transfer of emotions and ideas associated with an earlier relationship onto the analyst.

Ali's distance, his remoteness, was a quality that I had come to associate with the super-rich and celebrities. Perhaps the former are somehow benumbed by continuous exposure to exceptional experiences and the latter are discomfited by having to answer questions as themselves. Many celebrities are completely unable to step out of role and continue to behave like comedians, actors, or rock stars, even in the consulting room. This makes trying to help them almost impossible. It's like attempting to treat a cardboard cut-out rather than a real person.

Ali dressed very casually: ripped jeans and trainers, a

creased linen shirt – a cheap, hippy ornament hanging from his wrist. He wasn't a man who could be accused of flaunting his wealth. I noticed that when there were pauses in our conversation he very quickly looked bored.

He had been married to his wife, Yasmin, for almost twenty years and they had four children. It hadn't been an arranged marriage, as such, but their respective families had been business associates and both sides were keen for the couple to meet. The subsequent courtship was encouraged – perhaps even incentivised. Over time, Ali's uncles – who had been running his father's business – retired, and Ali became managing director. All of Ali's children attended private schools, and he and his family lived in a spacious house situated in a very desirable suburb. Ali had everything: inherited wealth, two sports cars and a devoted, beautiful wife. Then, one day, something happened which caused an unprecedented domestic upheaval. Purely by chance –she hadn't harboured any suspicions – Yasmin discovered that Ali had visited a prostitute. She had picked up Ali's phone with the intention of looking up a number in his contacts list, and was horrified when she came across a series of lewd text messages. This wasn't Ali's usual phone – it was his *other* phone.

Ali was slumped low in his armchair, almost supine – his legs fully extended. 'She was very upset. She wanted a divorce.'

'What made her change her mind?'

'I explained how stressed I was. It's not easy running a big business. And I've been doing it for a long time now. I explained that I'd been depressed for a while and that I wasn't thinking straight. She said: "If you're not well then you should get help and if you don't get help we're finished." So I said: "Sure – of course – anything."'

'What's making you depressed?'

He pursed his lips and didn't answer for several seconds. After such extended consideration I was expecting more than a single word. 'Stress . . . '

'How are you stressed?'

'There's a lot to do, you know? Responsibilities, admin; a few years back things were difficult for a while and I had to lay off a lot of people who were dealing with my paperwork. I had to start doing the book-keeping myself.'

'Couldn't you start employing people again?'

'Sure. It's something I should have done years ago. But I just never got round to it. There was always something else to do, something that needed urgent attention.'

'When you get stressed, how do you feel?'

'How do I feel?' I nodded. 'Well . . . not so good.'

'Do you get any symptoms?'

'Yes, I suppose so. Headaches.' He drew a line across his forehead with his finger. 'I get bad headaches.'

He wasn't very forthcoming. I asked him about his depression.

'When you get depressed, what kinds of thoughts do you have?'

'I think about the business – where it's going . . . '

'Anything else?'

'My marriage. I'm not proud of what I've done.'

His answers were consistently brief and uninformative. It was as though his lassitude made speech effortful. His fleshy eyelids drooped and he blinked – slowly – like a contented cat. He looked as if he was about to drift off to sleep.

Once, a patient of mine fell asleep while I was in the middle of a sentence. I knew that if I woke him up he would be embarrassed, so I waited until he roused and the instant he opened his eyes I finished my sentence. He was completely unaware that he'd been asleep for fifteen minutes.

I began to suspect that Ali had come to see me only because of his wife's threat. His illicit rendezvous had been exposed and he was now pretending to be unwell in order to prevent his wife from issuing divorce proceedings. I put this to him, phrased more diplomatically, and I was surprised by his reaction.

'No,' he said, raising himself up. He seemed genuinely disconcerted by my suggestion. 'I think Yasmin's right. I think I have got problems.'

'Do you think you could try to speak more openly about what's going through your mind?'

'Okay,' he rocked his head from side to side as if trying to free a trapped nerve. 'Okay. I'm not used to all this.' He referenced the generality of the consulting room with a lazy roll of his hand. 'I've never been a great talker.'

Our conversation remained insubstantial. Ali yawned, fiddled with the charm on his bracelet, and occasionally repeated himself: 'It's hard, running a business – a big responsibility. I'm not sure Yasmin appreciates that. She's never had to do it. I'm not saying that justifies what I did, no way – but still . . .'

'Are you resentful of her sometimes?'

'Resentful? Me? No, she's a great wife. Always has been; and a great mum.'

Perhaps I'd judged him too harshly. Perhaps he *did* have a significant problem but was simply too anxious to talk about it. Why was I thinking this? Ali didn't look particularly anxious and the tone of his speech hadn't changed. I just had a feeling – a hunch. I have already highlighted the perils of emotional reasoning. Wasn't I making the very same error that I warned my patients against?

In the late 1960s, the psychologist Paul Ekman undertook a study of depressed patients who were pretending to feel better. The reason for this pretence was that they wished to escape close supervision and attempt suicide. When filmed interviews with these patients were slowed down, it was possible to identify fleeting and negative facial expressions that were congruent with suicidal intention. In real time, these fleeting expressions would have lasted for a matter of milliseconds. It is possible that psychotherapists whose day-to-day work involves a great deal of looking at faces and trying to read them become sensitised to the appearance of negative micro-expressions. Such looks are perceived subliminally and produce a nebulous feeling that something is wrong. We know that this is possible from laboratory studies. Threatening images presented so quickly that the observer sees only a flicker of light can provoke physiological changes associated with fear (for example, increased sweat gland activity). What we call intuitions, presentiments and hunches might be nothing more than the by-product of pre-conscious processing: the registration of unconscious information that remains inaccessible to introspection.

We discussed Ali's marriage.

'Were you becoming sexually dissatisfied?'

'No. Sex with Yasmin is great. It's always been great.'

'Then why did you feel the need to visit a prostitute?'

'I don't know – I just . . .' He raised his hands and let them fall heavily on the chair arms. 'I just don't know.' Again, I got the feeling that I was looking at a very anxious individual. 'I guess I can be quite demanding,' he continued. 'I mean, sexually demanding. I need to have sex every day. Yasmin's okay with that, but I really do have a very strong sex drive.'

This small disclosure carried him a little closer to confession. His perspective changed and he recognised that a burden might be laid down. 'Even after we've had sex, I still feel that I need more. The need is so strong it wakes me up and I have to go to the bathroom to masturbate.'

'How often does this happen?'

'A lot – every night – sometimes several times a night – sometimes more . . .'

'So, on average, how many times do you ejaculate over a twenty-four-hour period?'

'About three I guess, although I don't produce very much.'

'Have you always been like this?'

'Yes, since I was a kid. In some ways I was worse back then; actually, a lot worse.'

There was a time when I would have been sceptical about such claims. Once, I would have supposed that a man approaching the age of forty would have neither the desire nor the capacity to orgasm three or more times a day – but not any more.

For a short period of time I worked almost exclusively with promiscuous gay men and rent boys, a truly humbling experience for a monogamous heterosexual. In the so-called 'fuck room' of a night club, a particular individual might have sex with ten or twenty partners and experience just as many orgasms. Pleasurable orgasmic sensations from the rectum and prostate are mediated by the pelvic and hypogastric nerves. There is a case report in the clinical literature of a man who experienced orgasmic sensations while defecating. Some individuals who suffer from epilepsy have orgasmic seizures that are triggered by brushing their teeth. One of my female patients could orgasm by popping bubble wrap – and she claimed that she could carry on having orgasms in this way indefinitely. For obvious reasons, I didn't ask for a demonstration, but I have to admit that curiosity almost got the better of me. It is always unwise to suppose that human sexual behaviour has limitations.

Ali's eyes, now wide open, suggested there was more to come. We had only minutes to spare. Ali rubbed his chin and said, 'I'm sorry, but I haven't been entirely honest.'

'Oh?'

'About the prostitute.' He rotated the charm. 'This is confidential, right?'

'Yes.'

'I mean – if I tell you the truth – you won't contact my wife.'

'No. Although that's something we'll have to discuss . . .'

'What? Whether I tell her or not?'

'Yes.'

'But *you* won't.'

'No.'

A pause – a ticking sound made with his tongue – and then a much longer pause.

'You were saying?' I said.

He stopped turning the charm and let it go. 'Yeah – the prostitute – you see – it wasn't one prostitute. I've actually been seeing prostitutes for a while now.' There was a sense of imminence, a sense of something coming. Ali continued: 'It's actually closer to three thousand – maybe more.'

'Three thousand,' I repeated.

'Yes.' Ali's expression was difficult to interpret: guilt and shame, macho pride, juvenile glee? All were represented in varying amounts.

'Is that possible?'

'I don't get up in the middle of the night just to masturbate. Sometimes I leave the house to have sex and come back again. Sometimes I have sex with several women a day.'

His wife's discovery of a single transgression had been Ali's admission ticket. Finally, he had volunteered his real problem – or so I thought. Actually, I was jumping to conclusions. This wasn't his real problem at all. His real problem was far more interesting.

The term 'sex addiction' first appeared in the 1970s. Prior to this decade, hypersexuality was described as nymphomania in women and satyriasis in men. Richard von Krafft-Ebing's notorious book of case studies, *Psychopathia Sexualis*, which was originally published in 1886, contains accounts of both: Case 193, a 'universally respected' farmer performed 'the sexual act from ten to fifteen times in twenty-four hours';

Case 186, a woman 'Of good ancestors, highly cultured, good natured, very modest', and who 'blushed easily' when left alone with a male of any age, 'would strip him naked and vehemently urge him to sate her lusts'.

Is it meaningful to talk about sexual needs as addictions? If the circumstances allow, the human animal – particularly the human male – is highly motivated to seek sexual satisfaction. Emperors, dictators, Hollywood actors and heart-throb musicians have all been known to have had thousands of sexual partners simply because they could. When rats can electrically stimulate the pleasure centres of their own brains by pressing a lever, they will do so indefinitely. Human beings, given the opportunity to have as many sexual encounters as they please, behave in much the same way.

Critics have suggested that the very idea of sex addiction is misconceived. They maintain that addiction is a term we should reserve for dependencies arising from the introduction of substances – such as cocaine, alcohol, or sugar – into the body. Behaviours such as sex, gambling, shopping, or playing computer games cannot be understood as addictions, because nothing is being ingested, injected, or absorbed. However, all behaviour has biochemical consequences and sex is associated with the production of endogenous compounds that resemble amphetamines and opiates. The adrenal and pituitary glands can manufacture hormones that deliver 'highs' and 'rushes' that are just as addictive as street drugs. As such, a restrictive definition of addiction that fails to recognise the reciprocal relationship between behaviour and biochemistry is somewhat arbitrary.

My own view is that sex addiction can be a useful concept

providing the individual's subjective state and context are taken into account.

Did Ali, for example, feel conflicted about his behaviour? Did he want to stop visiting prostitutes but feel driven to continue? Were there significant negative consequences for him and his family? Did he feel guilty? Over the course of our sessions, he answered affirmatively to all of these questions.

'Your wife never suspected?'

'No. She's not like that. It's not in her nature. She's got no idea.'

'She's a trusting person.'

'Yes. Very.'

'How do you feel about that?'

'Well, bad – of course. Look – I don't want to hurt her. Why would I want to do that? That's the last thing I'd want.'

'But if she knew the truth . . . '

'She'd be . . . ' The scene he imagined was too awful to articulate. He sat up and perhaps it was the slender hope of redemption that spurred him on. 'I don't want to be like this. I started going to see prostitutes when I was in my teens. I'd go to casinos with friends and one thing would lead to another. It's become more and more of a problem over the years – the whole thing has gotten completely out of hand. There's stuff I should be dealing with – people I should be seeing. And instead, I'm just out there.' He looked towards the window and the light lacquered his face with a silver sheen.

Addictive problems frequently escalate because of tolerance – a medical concept that can be employed to explain behavioural dysregulation and loss of control. The body has the capacity to adapt to drugs and sometimes larger and larger doses are required to achieve the same effects. In much the same way, the sex addict adapts to frequent sexual stimulation. Sex becomes less pleasurable, necessitating renewed efforts to reach an ever-receding goal. This effect is very pronounced in individuals addicted to internet pornography, many of whom spend hours or even days surfing increasingly exotic websites because ordinary sexual images no longer excite them. Libertines are often depicted in fiction as cynical, world weary and infinitely bored. It is a characterisation that acknowledges a paradoxical truth: the intemperate pursuit of pleasure makes pleasure hard to find.

Ali turned away from the window and continued talking as if there had been no interlude. 'I'm in a difficult situation right now. I've spent a lot of money . . . '

'You're in debt?'

'Yeah, you could say that.' Ali studied the floral pattern on the arm of his chair and traced around the edges of a tulip with his forefinger. 'Thing is, I think it's all gone too far now – I don't think I can come back from where I am.' I wasn't sure what he meant. Was he referring to his finances or his mental state? He saw my perplexity and added, 'The business. Things aren't looking good.'

'I'm not sure I understand . . . '

'Because of my habit.' Again he registered my perplexity. 'Some of the girls,' he continued, 'They're very expensive, you know, the pretty ones – the *really* pretty ones.'

'When are you going to tell your wife?' His mouth opened and his eyebrows climbed. It was my turn to clarify. 'Not about the escorts, about the business.'

'I don't know.'

'You must have been aware that the business was in trouble?'

'Yes. I do the books.'

'Couldn't you have taken preventative action?'

'I can't do anything now. There's no money left.' He pulled at his lower lip – let it go – and repeated the same words he'd offered his wife. 'I haven't been thinking straight.'

Sometimes, the reality of a situation makes us so anxious we simply pretend that it isn't happening. Among the many defence mechanisms identified since Freud's time, denial is one of the easiest to understand and it is also one of the most frequently observed. Everybody has some experience of denial. The defence is typically mobilised when symptoms appear which could be the first sign of a serious physical illness. 'It doesn't mean anything – it'll go away.' Some denial is helpful because it regulates exposure to bad news and facilitates gradual acceptance. The individual isn't overwhelmed. Extreme denial, however, precludes rational decision-making.

'It's going to be very difficult,' Ali continued. 'Yasmin's a proud woman and quite active in our community. You know, she does a lot of charity work – organising events, fundraising. Our family has a reputation.'

'I'm sure it will be difficult – but presumably, the collapse of your business will have consequences that you won't be able to hide from your wife for very much longer.'

He wasn't really listening to me. 'A man like me,' he said, nodding his head slowly, 'I'm supposed to provide. It's expected.'

Ali's business, established and developed over two generations, had been very profitable. I scribbled some rough figures in the margin of my notes. Even if Ali had paid for the services of over three thousand prostitutes, and a significant percentage of these liaisons had been with expensive escorts, his business should have been able to absorb the cost. Particularly given that his expenditure was spread over a period of approximately twenty years. I put this to him – and it was only then that he told me about his real problem.

'It's not just the sex. Things might have worked out if it was just the sex. You see, most of the money went on the other stuff.'

'The other stuff?'

He touched his wedding ring and then his charm. I wondered if Ali was performing a superstitious ritual, but didn't ask because I wanted him to stay focused. 'I'm not sure I can explain.'

I waited for him to continue but he remained silent. 'Try.'

Another extended pause – the sound of traffic – voices on the landing outside. 'All right, all right.' He allowed his head to roll back until his occiput found the support of a cushion. 'Usually, you pay for sex and walk away, but there are always some girls who get to you. I don't know what it is – how it happens – but it does.' He massaged his chest. 'I'd go back to see them and it would become a regular thing. And then it would become something more. We'd go on dates, like a real couple, and get to know each other. I'd take them out to nice restaurants; actually, very nice restaurants, the best, in fact, and I'd buy them gifts.' He grimaced before adding, 'That's where a lot of the money went – the other stuff.' His head was still tilted back and he seemed to be looking at an elevated point somewhere behind me. It was as though he couldn't make eye contact. He needed to divorce himself from the inhibitory influence of reality to complete his confession. 'With some of the girls, it got very involved. We'd make plans for the future. We'd chat about what our lives were going to be like, together. We'd go to see big houses, with an estate agent – and get really excited.' His eyes closed – and then opened again.

'So you did want to leave your wife . . . '

'No, never.'

'Then what were you doing? Why were you taking these women to view houses?'

'I've never wanted to leave Yasmin. It's like . . . with these escorts—' He broke off and I thought he was going to add, 'It was just a game.' But his lips came together and he set his jaw.

'How often did you do this?'

'Many times.'

'Yes, but how many?'

'I can't remember.'

I had become like an inquisitor, and the rapidity of my questioning had made him uneasy.

To enjoy fiction a reader must suspend disbelief and enter a world of fragile artifice. Any stylistic peculiarity has the potential to interfere with the reader's willing complicity. An awkward phrase, odd language or accidental innuendo can jolt the reader out of the story and make him or her aware that they are not observing the earth from space or hunting a great white whale but sitting in a room, reading a book. Psychotherapy, like fiction, also requires total immersion. Ali raised his head off the cushion and was frowning slightly. I had jolted him out of the 'flow' and it took me some time to regain his confidence.

The conversation that followed was halting and fragmentary. Even so, a clearer picture began to emerge. 'The sex was a means to an end, a way of getting closer. When I was younger, it was all about sex, but things changed. I wanted more. They're so beautiful, some escorts. You look at them – and it's like – sex isn't enough. You want more.'

Ali wasn't really addicted to sex. He was addicted to courtship. The candle-lit meals, the flowers, the extravagant gifts – necklaces, pearls and diamonds – eyes locked, finger tips touching across a table, the sound of violins, and a single rose on white napery – the elaborate ritual of romance. But there was still another layer to Ali's psychopathology, one more revelation, a final twist. 'It's not about me falling in love with them,' he said. 'It's really about them falling in love with me.' After the sex and the restaurants and the gifts and the shared fantasies and the house-viewing, his efforts would sometimes be rewarded by a declaration of love. This was what he craved most: the feeling of being fallen in love with. When he heard the words 'I love you' spoken with sincerity, the relationship had served its purpose. To achieve the same high again the entire process would have to be repeated with another woman.

'Did any of them try to find you?'

'I was always careful. I covered my tracks.'

The biochemistry of love is complex but we know that the rush of excitement when prospective lovers meet coincides with the release of phenylethylamine; sexual desire is stoked with testosterone; and feelings of attachment and blissful union are associated with oxytocin. When excited, dopamine – sometimes called the molecule of pleasure – spreads through the brain from the top of the brain stem, altering the responsivity of neurons in various regions, most notably the reward centres. I suspect that Ali was addicted to all of these endogenous 'drugs'. But it was the biochemical combination that underlies the feeling of being fallen in love with that gave him his biggest hit.

Addiction lends itself to reductionist analysis. Ideas such as chemical dependency and escalating need (driven by tolerance) seem very apposite; however, every addict is a unique individual. There will be numerous psychological factors, such as learning history, thinking style, impulsivity and vulnerability to mood disturbance, which will interact in complex ways. An understanding of both levels – the biochemical and psychological – enriches understanding of addictive behaviour and its maintenance. After Ali had made his ultimate revelation, I had countless questions.

Was his addiction to the opiate of 'being fallen in love with' a form of self-medication for pain that he was yet to disclose? Did he derive auxiliary pleasure from manipulating and deceiving his wife and the women who fell in love with him? How was he able to feel genuine affection for his wife while simultaneously betraying her on such a colossal scale? Was he a narcissist or a closet sadist? How did his problem develop over time?

At the end of our third session Ali got up and walked directly to the door. He paused and turned to look at me. I can only speculate, but he seemed to be making some form of evaluation. Had it been wise to reveal so much?

'Where are you going?' I asked.

'Right now?' He guessed what I was thinking and laughed. 'No – I'm not going to see a prostitute. I'm not that crazy.'

'See you next week?'

'Yes. See you next week.' He smiled and left the room.

Ali had started a relationship with me, consented to pay a fee for my services and made intimate disclosures that suggested I'd earned his trust. Yet, as soon as my concern for his welfare became explicit, he disappeared. He had treated me just like one of his escorts. A little worse, in fact, because he was a private patient and I never got paid. Ali didn't attend his next appointment and didn't respond to any telephone messages. Eventually his telephone numbers produced a continuous tone.

A few years earlier, at the end of a long day spent in a busy hospital outpatient clinic, I slumped down next to a colleague who was staring through his own reflection, out across a grim, urban landscape of rooftops and chimneys. It was obvious his day had been as arduous and emotionally draining as mine.

He raised his chin and said: 'What's the difference between going to see a psychotherapist and going to see a prostitute?'

'I don't know,' I replied. 'Tell me.'

'After seeing a prostitute at least one person feels better.'

Chapter 5

The Incurable Romantic:
On the impossibility of perfect love

When I was eleven years old I fell in love. This first seizure of romantic sentiment was preceded, almost a year earlier, by what I can only describe as a sexual awakening, although naturally I didn't think of it in those terms at the time. My mind was assailed by images of nudity associated with a shy, willowy girl in my class. These dim, fleshy intrusions were accompanied by physical sensations that I had previously associated with anxiety – quickening breath and butterflies in my stomach. As the physiological correlates of sexual arousal overlap significantly with those of fear, I found it all very confusing. The entire episode was probably attributable to my pituitary gland making some tentative secretions preparatory to my adolescence.

The girl who I fell in love with was called Susan. She was nothing like the shy girl from the previous year. Susan had blonde hair which she wore in a ponytail and there was something cutting and hard-edged about her. I had been aware of her for some time – but only as part of the general background noise of school. The first time she attracted my full attention was during an English lesson. The teacher was urging us to use the dictionary to improve our spelling and Susan quipped: 'If you can't spell a word how can you look it up?' I judged this remark to be outstandingly clever.

Labouring under a weight of inferiority, I found myself thinking more and more about Susan, and within a few days she had come to occupy my thoughts almost continuously. A spectral impression of blonde hair and blue eyes seemed to hover in the air in front of me, and the silence of my bedroom was broken by recollections of her voice. I began to feel miserable and wretched.

My journey home from school involved changing buses. I was waiting for the connection and felt so distracted and agitated that I had to start walking. It was a strange compulsion, obscurely motivated by a number of poorly defined and irrational objectives. Perhaps my agitation would diminish if I was in motion? Perhaps I might somehow travel beyond the boundary of my own misery? Perhaps the persistent image of Susan's face would fade? Whatever relief I'd hoped for didn't come about. Home was several miles away, so I walked and walked and walked and the rhythm of my step provided a beat over which I began to compose a song – a pathetic and inane lyric that expressed the utter hopelessness of my predicament. I can still remember some of the words and the simple melody after nearly fifty years. I would never find the courage to approach a girl like Susan, a girl with hair that flashed as brightly as her scintillating wit.

Eventually I got home. I did my homework, watched TV and went to bed. When I got up the following morning I felt a great deal better. Sitting in the classroom at school I was able to look at Susan and view her more objectively. Yes, she was pretty and I still wanted to talk to her, but the accompanying sensations and feelings were less intense. In fact, I didn't seem to be quite so in love. By the end of the week I was fully restored. The world seemed steady and balanced and normal again. Indeed, I couldn't understand why I had become so distracted.

Young love is problematic. A young person usually experiences falling in love for the first time during a period of accelerated growth that has no precedent other than the exponential division of cells that occurs after conception. The physical and psychological changes associated with adolescence – mature appearance, interest in sex, and brain developments – don't always proceed harmoniously. A particular individual might appear quite manly but think and manage emotions like a child. The prefrontal cortex – which mediates problem-solving, reasoning, planning and the control of emotions and behaviour in social situations – is the last part of the brain to develop. Indeed, the prefrontal cortex doesn't finish developing until an individual is in his or her twenties.

This is perhaps the single most important piece of information about children and adolescents that any parent needs to know. It explains a great deal. Most young people forget to do things, act impulsively, take stupid risks and make poor decisions, not because they are being wilfully provocative, but because their brains are unfinished. Moreover, in adolescence, the emotional volatility of the unfinished brain is made more extreme by sudden fluctuations in hormone levels.

Terms such as 'puppy love' and 'crush' trivialise the very real confusion, distress and hurt that the young can experience when they attempt to negotiate the emotional minefield of their first relationship. The consequences can be very damaging: sexual coercion, compliance because of peer pressure, abuse, regret, guilt, depression and, in some cases, rejection, heartbreak and suicide.

My first experience of falling in love was a preview of things to come. Even though it was brief, shallow and unconsummated, it contained many of the staple ingredients of a grand, romantic passion: obsession, idealisation, admiration from afar and the compulsion to channel feelings into some form of creative expression. Even my sudden urge to walk has innumerable precursors in romantic poetry and literature, which is overpopulated with heartbroken young men endlessly marching through varieties of indifferent landscape. Why was I behaving like a romantic stereotype at the age of eleven? How did I know what to do – even down to a certain degree of self-conscious parody? Some of my behaviour was undoubtedly programmed, but other aspects suggest that I had absorbed, unconsciously, models of behaviour by mere exposure to cultural influences. Romance has been characterised as the most significant belief system in the Western psyche. But what is romance? And what does it mean to be romantic?

*

'I just don't know why Imogen decided to end it. We were good together. We were happy.'

Paul's life had followed an ascending trajectory that reflected high expectations and achievements. He had been to a famous public school, studied philosophy, politics and economics at Oxford, and was subsequently offered a position in a private equity company.

'There were no signs, that's what was so shocking. She just said something like "I don't think we should see each other any more" and that was it.'

His delivery was flat – an inflected monotone – and his expression was curiously set. He was like a man talking through his own death mask.

'That was it.' He repeated.

Imogen's father was an art dealer and she worked in his central London gallery. It was while attending an exhibition at the gallery that Paul had met Imogen. He had been looking for some paintings to hang in his office – not for a new girlfriend.

'What were her reasons?' I asked.

'She didn't have any reasons. I asked her what the problem was, but she didn't explain herself very well. The best she could do was to say that she thought we wanted different things.'

'Why do *you* think she wanted to finish?'

'I honestly can't say. It's a mystery. We were very happy together and I can't believe I misread her.' He was an astute man and he anticipated my next question. 'There was no one else.'

'You're quite sure?'

'That's what she said and I have no reason to disbelieve her.'

His response was tetchy. It was as though my doubt had been ungentlemanly and he was bound to defend Imogen's honour.

I nodded, provisionally accepting his opinion. 'Sometimes people just drift apart . . .'

'We haven't changed.'

'And it isn't always easy to specify reasons.'

'We had something special. We didn't socialise a great deal, but when we did, my friends always made a point of saying how well matched we were. They could see it too.'

He'd stopped listening to me. He wanted answers but all that he could do was circle the problem, repeating expressions of disbelief. 'It's incomprehensible. I just don't understand how this happened. We were very happy together, I know we were. We didn't want different things. I never pressured her in any way, never made excessive demands. There was no need. How could she come to that conclusion? It doesn't add up.'

I thought it best to let him talk. His insistent repetitions were probably serving a purpose, helping him – I hoped – to come to terms with his situation. Eventually, he began to reminisce and he settled on the subject of Imogen's beauty. His praise became poetic. 'Sometimes, we'd be sitting in the same room, maybe on a Sunday morning, reading the newspapers, and I'd look up, and see her, and keep looking. I couldn't take my eyes off her, and, after a while, she'd notice, and say, "What's the matter?" and I'd say, "Oh nothing", and pretend to go back to the article I'd been reading. I never tired of looking at her.' He did much the same thing when she was asleep.

There is widespread agreement concerning the attractiveness of faces – a consistency of opinion that is observed across and between ethnicities and cultures. The orbitofrontal cortex seems to be the location in the brain where attractiveness is evaluated and the universal index of beauty is symmetry. A human face is highly complex and difficult to make. Therefore, when we look at a face, we are looking at the precise location where genetic mutations are most likely to find expression. Faces also declare the status of an individual's immune system. Our ancestors lived in an environment infested with disfiguring parasitic organisms. A symmetrical face and clear skin were reliable signs of resistance to infection. For ancestral humans, coupling with a beautiful mate was a good strategy for maintaining a stake in the gene pool. It still is. Handsome men have better sperm quality, and a comparable study of beautiful women would no doubt reveal a superabundance of viable eggs. Individuals with symmetrical features also smell nicer to the opposite sex.

Psychologists and sexologists frequently refer to the Coolidge effect, a term that acknowledges the power of novelty to incite desire while simultaneously celebrating the dry humour of the thirtieth president of the United States. The story goes that Mr Coolidge and the First Lady were being shown around a farm on separate tours. When Mrs Coolidge observed a rooster mating she asked how frequently the bird managed to perform. The guide said that it was dozens of times a day. 'Tell that to the President,' said Mrs Coolidge. Her message was dutifully conveyed to the President, who enquired: 'Same hen every time?' The guide shook his head. No, the rooster mated with many hens. To which the satisfied President responded: 'Tell that to Mrs Coolidge.'

Paul and Imogen had been together for only four months. Novelty is a powerful aphrodisiac, even more so when it is compounded by beauty. The sudden withdrawal of the object of Paul's desire, at a time when he was still in the first, heady throes of love, had made him profoundly depressed. There was something about him that reminded me of an addict, a shivering enfeeblement.

He changed position, slowly and with evident discomfort, as if every muscle in his body was causing him pain, and said: 'I'll never meet anyone like her again.'

The sentence was spoken with obstinate conviction.

Paul's relationship history was unremarkable for someone of his class and age, and he had always dealt with rejection calmly and philosophically. There were plenty of single, intelligent and beautiful women in his social circle and many of them seemed to congregate at dinner parties thrown by his friends. He had once dated a famous actress renowned for her stunning looks. There was nothing obviously different about Imogen. In many ways, she was a facsimile of all his other girlfriends.

I asked Paul to reconsider. 'Really? You'll never meet anyone like Imogen again?' He reminded me of her beauty and counted off ancillary feminine virtues using his thumb and fingers like an abacus. When he had used up all of his digits he flicked his hand, communicating a certain disdain for mathematical reductionism. 'But she had something else – something more – something I can't put into words.'

Atheists often attack believers for invoking the 'God of the gaps'. Whenever a gap in scientific knowledge is identified, the gap is enlisted as a religious proof. We cannot explain the ultimate origin of the universe. Therefore, God the creator must exist. We make the same thinking error when we are in love. The reasons why we fall in love with one person rather than another are simply too numerous, subtle, elusive and complicated to untangle. Many of them are the result of unconscious processes. Subsequently, there are always gaps in our understanding of love and like hopeful theists we tend to fill those gaps with supernatural explanations. We hint at strange affinities and the operation of mysterious forces.

'She was unique.'

'Isn't everybody unique?'

'She stood out from the crowd. She really did.'

Imogen was beautiful. But she wasn't beautiful in the same way that other women are beautiful. Her beauty was of a different order, a different magnitude. She was beautiful in the same way that a fairytale princess is. She was surrounded by swarms of glittering motes and encircled by rainbows.

Psychoanalysts view idealisation as a defence. It simplifies the world in order to reduce the anxiety caused by inconsistency and troublesome complexities. Idealisation always incorporates a degree of denial, because in order to see someone as perfect we must deny the existence of their less favourable attributes. At some level we must 'split' them into two and ignore the half we don't like. The term 'splitting' is most strongly associated with Melanie Klein – one of the first psychoanalysts to specialise in the treatment of children. She achieved this by introducing toys into the consulting room and interpreting play.

For Klein, the origins of splitting date back to the early weeks of existence, when the most significant event in a baby's life is feeding and a mother is not perceived as a person but a pair of breasts. Sometimes the milk flows, resulting in feelings of blissful contentment (a state that presages the ecstasies of being in love), and sometimes the milk is less plentiful, or even worse, absent, which results in feelings of frustration and anger. The baby does not understand that the good, bountiful breast and the bad, empty breast are both attributes of a single person. Eventually, the baby develops the intellectual capacity to recognise and accept the truth. The good breast and the bad breast are part of his or her mother – an inconvenient, refractory and consternating mix of good and bad.

The ability to assimilate the complexity of others is a measure of maturity and a necessary condition for the formation of authentic, meaningful relationships. Some people, when lovesick, revert to the primitive, and continue to manage their anxieties by splitting. The good is embraced and the bad denied. In this way, a besotted man might distort reality and perceive an ordinary woman as a goddess.

'Did you and Imogen ever argue?' I asked.

Paul looked around the room as if the answer to my question had been written on one of the walls. Finally, he answered: 'No, not really.'

'When you say "not really" . . .'

'We disagreed sometimes. We used to discuss lots of things – politics, paintings, music. She had strong opinions about modern art, some of which I found quite extreme. But I loved her passion.'

'While you were together, Imogen never said anything that upset you?'

He considered my question for a few moments. 'She . . .' But he did not finish the sentence. He was unable to say anything against her. The very idea of criticising her made him feel uncomfortable. He coughed into his fist, embarrassed and suddenly wary.

The notion of idealisation existed long before psychoanalysis. In the eleventh century, the Persian physician Avicenna considered it to be a key symptom of lovesickness. He attempted to persuade patients to be more realistic about their objects of desire by challenging their beliefs – a surprising commonality with contemporary cognitive therapy. For male cases of lovesickness, he also advocated exposure to articles of the beloved's clothing smeared with menstrual blood. The purpose of this was to force the patient to acknowledge the corporeal reality of the beloved and devalue the idealised mental image. He was, in effect, attempting to provoke the therapeutic assimilation of overvalued and denied parts.

I continued testing the strength of Paul's idealisation with gentle challenges but Imogen had been raised on a very high pedestal. My efforts simply glanced off Paul's defences.

'I love her so much.' He placed his elbow on the arm of the chair and supported his head with his hand. There was something affected about this pose. He seemed to be imitating Hamlet or Byron. 'She *must* feel something for me . . .'

'Why do you say that?'

'She must.'

'Because you love her . . .'

'We understood each other.'

'I appreciate how painful this is for you but sometimes love – however deep – just isn't returned.'

He raised his head from his hand and glared at me as though I had uttered a blasphemy. I had contradicted a sacred principle: if you love someone enough, your love will be reciprocated. It is an instance of a more generalised assumption that social psychologists call the just-world hypothesis: You deserve what you get, and you get what you deserve. The world, however, is not a fair place. There are no invisible forces at work restoring moral equity, and sincerity of feeling has never guaranteed that a declaration of love will be accepted.

I held his censorious gaze. 'What are you thinking?' I asked.

His expression softened and his eyes dimmed. 'I felt so close to her, closer than I've ever felt to anyone before. After we slept together for the first time, I was holding her in my arms and I can remember thinking, this feels so right. Our bodies just seemed to fit together.'

According to a Greek creation myth recounted by Plato, human beings were once double-headed creatures with eight limbs and represented by three genders: male, female and hermaphrodite. Zeus' punishment for human pride was to slice every individual into two equal parts, producing the human bipedal form with which we are all familiar. The legacy of this historic retribution is a pervasive feeling of incompletion that we suspect can be negated only by reunification with our lost half. It is a narrative that accounts for our deepest romantic yearnings while providing a neat explanation of homosexuality, heterosexuality and lesbianism. Plato's myth demonstrates that the deep satisfaction described by Paul, the sense of becoming whole again through sex and holding, unites modern and ancient civilisations. The desire for completion is perhaps finally mitigated when myth and evolutionary objectives coincide with the production of children. Thereafter, the wound inflicted by Zeus is healed and yearning is replaced by more practical necessities such as paying bills, doing housework, getting the kids to school and trying to get a good night's sleep.

'I'm sure she still loves me,' Paul continued. 'I think she's just confused. I'm sure that deep down – somewhere – in her heart – there's still something . . . a connection. Perhaps it was all too intense. You know, too much – too soon. And she felt overwhelmed. That happens, doesn't it?'

'You're sure she still loves you. But how do you know that?' He offered me a few non sequiturs but I persisted. 'How do you know what Imogen's thinking?'

'Well . . . we're on the same wavelength. Anyway, that's what it feels like.'

I pressed him on this point. A patient suffering from de Clérambault's syndrome would assert that he or she possessed certain knowledge of their beloved's thoughts. Fortunately, it transpired that Paul wasn't delusional. 'I can't say for certain,' he confessed. 'Of course I can't. But when you know someone well, you can make educated guesses.' Lifting a rigid forefinger, he wagged it in my direction. 'I do know one thing for certain though. I can make her happy. If she just gave me a little more time – one more chance . . .'

'She was quite clear, wasn't she? She doesn't want the relationship to continue.'

I had been too direct, too insensitive, and my unqualified words made him clutch for any means of salvation. A broader, philosophical view offered him new hope. 'Maybe this has all happened for a reason. It's said that when bad things happen, they make us stronger.'

'Nietzsche,' I said – then instantly regretted it. There was no need to identify the provenance of this maxim other than to satisfy my own intellectual vanity.

I was once having lunch in a hospital canteen with an ageing and very distinguished psychiatrist. He was so old and well known that he'd already had a mental health centre named after him and I'd seen him portrayed in an award-winning film. Yet, he was always unassuming, solicitous and extremely modest. He had just come from a meeting with an in-patient and her husband. 'Oh dear,' he said. 'That didn't go quite the way I'd hoped. She was very depressed today so I thought I'd make some helpful and inspiring analogies. I started to talk about the Battle of Britain and Churchill – and then the Allied invasion of Sicily – and I got completely carried away. The poor woman thought I'd gone mad. The husband was speechless.' Whenever I remember comparable transgressions of my own I console myself with this memory.

'What?' Paul's head jutted forward.

'Nietzsche,' I repeated. 'The philosopher . . . '

'Yes,' Paul replied with sour irritation, 'I know who Nietzsche is.'

'That which does not kill us makes us stronger.'

'Yes, yes . . .' The quotation seemed to revive his spirits and he sat up. 'If we did get back together, I'd be a stronger person – a better person. And all this misery would have had a point – a purpose.'

He was reassigning a different meaning to his separation from Imogen. It wasn't an end, but a beginning – a trial that he must endure to win true and supreme love. His thinking was following a convention of courtly romance. Like an exiled Arthurian knight, he would face temptations and dangers, pass tests and return in triumph providing evidence of his virtue. 'If we did get back together again, I'm sure we'd appreciate each other more.' He was going to demonstrate the durability of his love and win his Queen's favour. 'It might be better – second time round.'

The sensation of descent in my chest confirmed the accuracy of a good metaphor: my heart was sinking. It isn't by accident that the words 'incurable' and 'romantic' are paired with such frequency. To say one almost immediately suggests the other. Clichés can be very informative, which was why – even at this early stage in Paul's therapy – I was not optimistic about the outcome.

The next time I saw Paul, he was looking dishevelled. He was pale and had neglected to shave.

'It's difficult, without her.'

'What do you miss most?' I asked.

Paul tangled his fingers – an awkward interlacing that suggested deformity. 'I don't want you to get the wrong idea about this, but I'm going to say the sex. Because sex with Imogen was never just sex, it always felt like something more. It felt . . . ' The gap between his eyebrows narrowed as he struggled to overcome the limitations of English. 'This will sound ridiculous, but it felt . . . out of this world – outside of time. We would spend whole weekends in bed. And it was like nothing else existed.'

'Do you think about death much?' I asked.

Paul was surprised by my question, but was quick to recognise its pertinence. He smiled and made a noise – a soft grunt. 'As a matter of fact, I do.' He managed a gravelly chuckle. 'Even when I was a child I was quite morbid. My mother used to say to me "it's a long way off, nothing to worry about at your age". But I wasn't reassured. I could recognise a platitude even then.'

'Do your family have a faith?'

'No. My mother and father are both atheists.'

'And you've never found any comfort in religious teachings yourself?'

'No – it's all so implausible. Religion has never held any answers for me, which is a shame, because, actually, I'd like to believe in something . . . '

But he did believe in something. He believed in love.

The Greek philosopher Epicurus maintained that all of our anxieties and sadness can be traced back to a root cause: the fear of death. It is a view that has found considerable currency among practitioners of existential psychotherapy – a post-Freudian school that became increasingly influential throughout the 1940s and 1950s. A great deal of existential psychotherapy is concerned with the search for meaning, which must be personal, because the universe is intrinsically meaningless. We must decide what is meaningful for ourselves.

Love gives us purpose. And sex, which promises a surrogate form of eternal life through procreation, reduces (albeit temporarily) the potency of the two great existential terrors – aloneness and mortality. Sexual union anaesthetises the pain of solitude, and the psychoactive substances released into the blood when we are aroused can take us out of time and make us feel that we are boundless and eternal. In the ecstatic delirium of orgasm we are beyond Death's reach. Paul needed Imogen to be perfect because her perfection was protective. Love made him immortal.

Shortly after my sixteenth birthday I was sitting in the classroom of a further-education college listening to a lecturer reading Dylan Thomas's *Poem in October* which begins: 'It was my thirtieth year to heaven.' When the lecturer had finished reading he asked me a question: 'Why thirtieth year? What's so significant about that?' I didn't know. I didn't really understand the poem. 'Well,' said the lecturer, 'thirty is the age when you have to accept – beyond any doubt – that you've got one foot in the grave. It's the point in life where you realise that death is non-negotiable.'

Paul was thirty-one.

'I did something that didn't really work out.'

'What did you do?'

'I gave Imogen a call.' His lips became bloodless as he pressed them together. A few seconds passed and he continued: 'I wanted to see how she felt about things. It's been a few weeks now and I thought she might have moved on a little, you know, be more willing to talk about what happened.' He shook his head – an evanescent tremble. 'She wasn't interested. She said she was sorry that I was unhappy but she had nothing else to say. I tried to keep the conversation going, asked her to tell me what I'd done wrong, how I could make things right . . . ' Paul picked at a cuticle on one of his fingers. 'After I'd put the phone down, I felt annoyed at myself.'

'Why?'

'I hadn't really told her how I felt. I'd wanted to sound calm, reasonable – but it was all play acting, insincere. So I phoned her again.'

'How much later?'

'Not long – ten minutes. Fifteen maybe.'

'Okay.'

'I poured my heart out. I told her that I loved her and would do anything to have her back in my life. I begged her to give me another chance.' He swallowed and the prominence of his Adam's apple moved up and down. He was finding it difficult to say his next sentence.

'And how did she react?'

'She said that she didn't want me to call her – ever again.'

'That must have been very hard for you.'

Paul began to take deeper breaths and when he spoke again his voice was catching. 'Talking to her, knowing she was there – at the other end of the line – and that I might never . . . I love her so much.' His head fell forward into his hands. At first, his agony was silent, but his attempts to contain his emotions were futile and intermittent vocalisations became loud sobs. 'Shit,' he said. 'I'm sorry about this.'

'There's nothing to apologise for.'

Paul looked up. The flesh around his eyes had become swollen.

'Does this happen a lot?'

'Yes.' I handed him the tissues.

'Thank you.' He wiped away the tears and blew his nose.

'All right,' I said. 'Perhaps you could tell . . . '

Paul raised his hand and interrupted me. 'No, no . . . there's more.'

'Okay . . .'

'I kept on going over what we'd said to each other – sentence by sentence – and I came to the conclusion that trying to talk to Imogen on the phone had been a bad idea. I convinced myself that if I'd spoken to her in person we might have got somewhere. So the next day, which was Saturday, I drove over to her flat.'

'You didn't let her know beforehand?'

'No.' Paul dabbed his eyes again with the tissue. 'Her flat has an entry phone and a security camera. She was annoyed and told me to go away – but I rang again and she let me in. She was waiting for me outside the lift. She told me I was frightening her and I said, "Don't be ridiculous". How could she be frightened of me?' Actually, it was quite easy to see how she might be frightened of him. His desperation created an impression of instability. '"Do this again," she said, "and I'll call the police." She walked away and I followed her to her door. She slammed it in my face.' He flinched – re-living the moment. 'I know it looks bad – like I'm harassing her – bothering her – but all I want is a chance to talk, that's all.'

'She *has* been very clear. She wants you to leave her alone.'

Paul looked down at his tissue, leaned over the chair arm, and dropped the crushed ball into the waste paper basket. 'It feels wrong to just let her go. I mean, there are so many songs and films – and they all have the same message: love will find a way, love conquers everything.'

'They're pop songs.' I exploited the plosive properties of the word pop. 'Hollywood movies . . .'

'Sure, sure, but that's what *we* believe. That's why they're popular. They strike a chord.' He suddenly looked bashful. 'I wrote a poem last night. I haven't written a poem since I was at school.'

'Did it help?'

'Yes – I think it did. Putting my feelings into words . . .'

'Do you want me to read it?'

He smiled. 'God no.'

Evolutionary psychologists have suggested that artistic expression is a male fitness indicator. Skilful behaviours such as singing a song, painting a cave wall, or telling a good story advertise good genes. Moreover, the emotional roller-coaster of love, with its many ups and downs, mimics the oscillating mood disturbance associated with artistic talent. It is as though falling in love optimises an individual's creative potential so that they are viewed in the most flattering light by potential mates.

I have seen many couples for marital therapy and the most frequently aired complaint was voiced exclusively by women: he doesn't talk about how he feels. Men are famously uncommunicative and lacking in emotional intelligence. Yet, when I asked these wives what their husbands were like at the very beginning of their relationships, it was a different story: love letters, telephone calls, pillow talk – occasionally poems and songs. Falling in love loosens a man's tongue. However, when a man becomes eloquent, women should note that his lyricism will last only as long as it takes him to ensure the survival of his genes.

If – as many believe – creative expression was selected to ornament the male courtship display, it does *not* follow that women are intellectually inferior. For a display to be competitive it must be understood and valued. Without a discerning audience, all displays become wasteful. Nor does it follow that women are less gifted than men, although it is probably true that they are less inclined to advertise their creative achievements quite so loudly.

'You won't try to see her again – will you?'

'No.'

'Or call her?'

'No.'

'Because if you do . . . '

'Yes, yes. I understand. I'm not going to.'

We talked about his future, the prospect of forming other relationships. But he was reluctant to consider this possibility. Even so, I thought it beneficial to float the idea, to prepare the ground for subsequent conversations, because at some point he would have to start thinking about letting go of Imogen and transferring his affection to someone else.

In reality, few people get to marry their ideal partner. Love involves making a series of compromises. This is no bad thing, because an idealised partner is only nominally human.

'In a sense,' I said to Paul, 'the woman who you want to talk to isn't really there any more. Perhaps she never was.'

He considered what I'd said and shrugged. 'To be honest, that doesn't make it any easier.'

One week later, the door opened and when Paul entered I was immediately struck by his feverish appearance. He dispensed with the usual courtesies and blurted out, 'Something terrible has happened.'

'Please . . . ' I gestured towards the chair.

He sat down and his agitation made his fingers writhe. 'I didn't call her, just like we agreed.' He said this as though I had accused him of breaking his promise. 'But we sort of met – by chance.' His mouth warped out of shape and he added a qualification. 'Well, that's not strictly true.' He tried to calm himself by inhaling and letting the air escape slowly. 'You see, I was driving along and I saw her getting into a cab. I didn't stop, I just drove past, but I was aware of the cab – behind me – I could see it in my mirror. Anyway, the cab caught up at some traffic lights and I saw Imogen in the back and I thought: this is odd.'

'Why odd?'

'Well, what are the chances of that happening – in a city the size of London?'

'More than you might think.' I reminded him that human beings frequently misjudge probabilities.

'I was hardly going to get my calculator out . . . '

'No. But you invested this chance event with special meaning.'

'Yes, I suppose so.'

'Your paths had crossed for a reason.'

He didn't want to discuss his misattributions and was anxious to make further disclosures. 'When the lights turned green I let the cab go ahead and I just – well – started to follow it – until it pulled over outside Imogen's.'

'Just a moment, why were you in her area?'

'Well, I wasn't – not initially.'

'So how long were you following the cab for?'

'Not *that* long – twenty minutes? Anyway, I parked and she saw me get out of my car and as I walked towards her she went crazy. She swore at me and told me to leave her alone – and when I tried to explain what had happened – she ran off.'

I was about to ask him another question when Paul – as in the previous session – held up his hand and said, 'No, no – there's more.'

'Okay.'

'She called the police.'

'Okay.'

'They came round to my place and cautioned me. But I wasn't stalking her – I . . . I wouldn't.'

'You know if you don't stop bothering her . . . '

'Yes, yes – it's going to end badly.'

'What if this happens again? What if you see her in a shop or in a bar? What will you do?'

'I'll turn around and walk the other way.'

'Will you? Or will you think, this is odd, and conclude that fate is drawing you together again for a purpose?'

He nodded – a rare instance of acquiescence. But there would have to be many more instances of insight, otherwise, I feared, Paul might end up swapping his penthouse for a prison cell.

The word 'romantic' is extraordinarily rich and complex, because it represents many beliefs and ideas about love that have accumulated and blended together over a period of a thousand years. The concept of romance is so much a part of our cultural heritage that we accept its implicit assumptions without question. In plays, operas, films and novels, anything – if it is done for love – is acceptable.

Today, Islam is frequently characterised as an exporter of hate; however, in actuality, the Islamic world's most successful export is love. Our concept of romance has a Middle Eastern pedigree. The Arab Bedouin composed a form of poetry that contained several motifs now familiar to a global readership: an idealised lover, thwarted passions and melancholic yearning. Building on this tradition, eleventh-century Islamic authors wrote large-scale epic romances. The dissemination of Islamic love stories across Europe followed the Moorish conquest of the Iberian Peninsula. Presumably, these stories were told and retold by travellers who had crossed the Pyrenees and in due course they were cannibalised by the itinerant entertainers of medieval France. Thereafter, the chivalric songs and verses of the troubadours provided the foundation for home-grown courtly adventures featuring radiant queens and 'beautiful ladies without mercy' whose remoteness inflamed knightly passions. During the Renaissance, poets such as Petrarch and Dante took the theme of idealisation to new and ecstatic heights. The word 'romance' was infused with further meanings in the late eighteenth century, when romanticism – a movement that valued violent passions over cold reason – found its initial impetus in a story of doomed love by Goethe: *The Sorrows of Young Werther.* This slim volume, which ends with the protagonist committing suicide, was massively influential and forged a strong link in the popular imagination between love and death. Numerous imitators glorified the misery of rejected lovers in poems that depicted young men setting off across winter landscapes with suicidal intent.

The fundamental problem with the notion of romantic love is that it is based on a misunderstanding. Early Islamic romances were allegorical and dramatised the soul's longing for God. They were never intended to be taken literally. By confusing spiritual and earthly goals, Western authors imported a raft of unrealistic expectations into courtship and marriage. How can a mortal woman ever live up to the romantic ideal of eternal beauty? How can an imperfect human being deliver perfect love? Is it really the case that there is only one person (like a singular deity) with whom true love is possible? Sex, however pleasurable, is not heavenly communion. Fate (or the hand of God) does not bring people together, there are only random occurrences. Obstacles to love have no significance; they do not appear in order to test and intensify love. There is no divine plan.

Romantic love makes impossible demands and quickly falls apart, after which its wretched, disappointed devotees are offered the cruel consolation of a freezing landscape and a pistol. The romantic world view is rooted in literatures that construe love – particularly young love – as nascent tragedy. As such, it is a potentially dangerous body of ideas.

To be romantic is, for the most part, an unhappy, hallucinatory experience. Romantic love promises one thing, but delivers another.

The trappings and accessories of romantic love have now been successfully monetised. On Valentine's Day we celebrate romance with cards, bouquets and candle-lit meals, and presents of chocolates and lingerie tastefully packaged in red ribbons and love-heart gift-wrap. But what are we celebrating, exactly?

The French psychoanalyst Jacques Lacan – a pupil of de Clérambault and something of an intellectual playboy – asserted that one of the most significant milestones in psychological development occurs when an infant sees him- or herself in a mirror for the first time. Self-recognition is followed by the disconcerting realisation that what others see of our exterior form does not correspond with our more vital, fluid and authentic inner world. All mature adults must accept that they are essentially unknowable – and that they will never know the one they love. Even when we kiss there is distance; it is a distance that cannot be bridged by romantic love and must be respected if a relationship is to succeed. The real metric by which we can gauge the authenticity of love is not how close we want to be, how merged and intermingled, but how far we can stand apart and still be together.

'Did anything happen in your childhood that made you especially aware of death?'

Paul's expression was neutral. 'No. Nothing.'

'A death in the family?' He shook his head. 'The death of someone at school?'

'Certainly not.'

'Pets . . . '

'We didn't keep pets.' He raised his hands and let them fall. 'It's just the way I am – there's no accounting for it really. When I was a kid, the thought of dying terrified me. I used to get this horrible feeling in my stomach – dread, I suppose. Now it's more to do with pointlessness. If we're all going to die, what's the point of anything?'

'Some people think the exact opposite. It's precisely because things are transient that they have greater value.'

'Really? Not me.'

We discussed how pursuing perfect love gave him purpose and temporary relief from existential angst. He was interested in these ideas and listened intently. I suggested that if he could accept death with greater equanimity he might feel less compelled to seek solace in romantic idealism.

It is natural to fear death; however, for some people this fear becomes so intense and troublesome that they can't enjoy life. When this happens, the condition attracts medical appellations such as 'death anxiety' and 'thanatophobia'. There are a number of arguments that can be employed to help people with death anxiety. They aren't always effective, but when they do work, patients experience a change of perspective and the idea of death becomes less alien and strange.

We are more intimate with oblivion than is generally acknowledged. Every night there are discontinuities in our existence during dreamless sleep. Moreover, we forget things every day, so in a sense we are constantly dissolving into nothingness. Recognising that our nativity was preceded by aeons of oblivion can, for some individuals, turn 'the great unknown' into 'more of the same'. The chemical constituents of our bodies were assembled by exploding stars at an inconceivably distant point in the past, and these constituents continue to exist, in some shape or form, after our death. We are woven into the fabric of the universe – and will always be. There is also a kind of afterlife in procreation, making cultural contributions, leaving legacies, or simply being remembered by those who survive us. By merely existing, we influence an expanding web of cause and effect relationships that will continue indefinitely.

Freud claimed that none of us really believe in our own death. Although this is probably true when we are young, it undoubtedly becomes less true as we grow older. Paul had reached a point in his life where he could no longer deny his own mortality. I suspect that if Imogen hadn't appeared, then any attractive woman would have been cast in the same role. His love had much more to do with what he wanted her to be, rather than what she actually was.

'I'm not sure life is worth living without her.'

I asked him directly about whether he was having suicidal ideas.

'I've thought about ending it all, but only in an abstract way. I mean, I haven't considered how I'd do it.'

'You said life isn't worth living without her . . . '

'Yes. That's what it *feels* like.' He became tearful and I was curiously reassured by this. Suicide risk is more frequently associated with emotional numbing. It's as though suicidal patients are too sad to cry. 'I don't want to die,' he continued. 'I want to live – but I want to live with her.'

Paul was excavating romanticism all the way down to its spiritual foundations. Imogen had become his everything, and the light in her eyes emanated from the scented gardens and fountained courtyards of a Middle Eastern paradise.

Week after week, Paul came to my consulting room to express his longing. Sometimes, I would simply listen, and at other times (particularly if he was looking stronger) I would point out how his unhappiness was being maintained by a belief system full of contradictions and dysfunctional assumptions. A few hairline cracks began to spread across Imogen's idealised image. Paul was prepared to admit that she wasn't always very reliable.

'When people are consistently late, what might that mean?'

'Perhaps they're busy.'

'Are all busy people consistently late?'

'No.'

'So what else might it mean?'

'They could be disorganised.'

Sometimes I had to abandon Socratic questioning. 'Or maybe they just think *their* time is more important than *yours?*'

'You think she was selfish . . .'

I let the last word resonate before I spoke again.

On my way to work I crossed a park full of colourful flowers. The season had changed.

Paul was looking better.

'Are you ready to start dating again?'

'Not yet. But soon . . .'

'And what are the chances of you meeting someone else – someone who you can love again?'

He linked his fingers and bowed his head – an attitude that suggested prayer. 'The honest answer is I don't know. It might happen.'

At least he was admitting the possibility of life after Imogen.

Paul scratched the back of his neck. 'I've decided to work abroad. An opportunity has come up – in the States.'

'That's rather sudden.'

'Not really. I've been thinking about going to America for a while now.'

His announcement made me feel uncomfortable. 'Are you sure that you haven't made this decision because of Imogen?'

'I thought it might be the right thing to do ... make a new start.'

'Actually, I was thinking that maybe you don't trust yourself.'

'I'm not going to pay her any more surprise visits.'

'You certainly won't if you're living in America.'

My remark was perhaps a little too pointed.

I saw Paul for another two sessions. We reviewed our key insights and discussed whether or not he should continue seeing another therapist in Chicago. 'I'll see how I feel,' he said. 'And take it from there.' He shook my hand, thanked me for my help and said, 'It's odd – this. I've told you so much. You know so much about me and I know almost nothing about you.'

'What do you want to know?'

'Presumably ... you've been in love?'

'Yes.'

'Good.'

We both laughed – and then he left.

Letting go of patients is a peculiar thing. For me, those last moments are always accompanied by a very particular sadness.

A year or so after our final session I received a letter from Paul. The content was pleasant and good humoured although somewhat superficial. He had come across one of my novels and enjoyed it. His career was flourishing, the business environment in America being generally more favourable for venture capitalists. As I read the letter, I became increasingly aware of my own expectancy. I read faster in order to get to the paragraph where he would tell me that he was happy and in a new relationship. But there was no such paragraph. Rather foolishly, I turned the sheet over and scanned both sides again. I found nothing to dispel my disappointment.

How ironic that I should want a happy ending: lovers walking off into a sunset to the sound of a thousand, soaring violins. How absurd that I should harbour a desire for Paul's life to reflect the formulas of romantic fiction. I folded the letter, slid it back into the envelope, and put it away in my desk. The power of romance should never be underestimated.

Chapter 6

The American Evangelist: *Sins of the flesh*

In my mid-twenties, I left London with my wife and six-month-old son to live in a remote village in the north of England. My wife and I had met at a further-education college when we were sixteen and seventeen, respectively. We both came from working-class backgrounds and neither of us had been offered much in the way of guidance. Although I gained a few qualifications at college, I didn't apply to go to university. No one in my family had ever been to university – my mother and father completed their compulsory education at fourteen and left school at the earliest opportunity. University was something other people did. Luckily, I had been taught to play the piano by a relative and I was able to earn a modest income giving piano lessons to children.

I couldn't get any pupils in the village because there was no demand. I entertained the idea of becoming a writer but this was completely unrealistic at the time. My wife had no plans other than to do some occasional bar work in the nearest market town. We were living on benefits and we had, in effect, dropped out. Why had we chosen to live this way? I could advance some reasons that might elicit sympathy, but the truth of the matter is that we were immature, irresponsible and stupid.

One day was much like the next. The sun came up and the sun went down. We couldn't afford books but a mobile library occasionally visited the village. I read, listened to the radio and went for walks – pushing my son in his buggy.

In spite of our impecunious circumstances my wife and I were happy. The decision to leave London had been a joint one – our thinking strongly influenced by the fashionable escapism of the time. We were (it goes without saying) inexcusably naïve.

The principal attraction of the village was its romantic and austere location. Out of my kitchen window, I could see stone dwellings protected by an amphitheatre of higher ground. And beyond the village, in all directions, there were rugged fells, fields, rivers, ruined castles and moorland. It was a very evocative landscape, steeped in Arthurian legend. One of the local ruins was called the Dolorous Tower.

Behind our cottage was a hill surmounted by an eleventh-century church. It was a striking building. The bell tower had a parapet with distinct pinnacles. For generations, the church had been associated with a folktale that I found darkly appealing. Originally, the villagers had wanted to build their church at the foot of the hill, but every night, when the labourers had gone home, the stones and timbers were mysteriously transported to the summit. All agreed that this was the Devil's work; his purpose being to force the villagers to build their place of worship in an inaccessible location. The Devil proved to be an indefatigable adversary and in due course the villagers conceded defeat. It was an unusual conclusion to a folktale. Folktales are instructive and the Devil is usually outwitted, evil is overcome and good prevails. This story, however, had no reassuring moral. Satan was triumphant.

The church was damp and cold, and the air inside imbued with the mildewy fragrance of rotting missals. I would climb the hill, let myself in and play the ancient, wheezing harmonium. It was quite eerie, sitting alone in that church, and against my better judgement I frequently found myself looking over my shoulder – disturbed by imaginings. Eventually, I discovered an Edwardian history of the village and read that the hill had been inhabited long before the eleventh century. It was the site of pagan sacrifices prior to the arrival of Christianity.

As usual, I was attracted to the sinister – the legends, the strangeness – and it hadn't escaped my notice that my situation was a cliché favoured by horror writers. A young couple go to live in an isolated location, foolishly cutting themselves off from friends and family. They have a small child – a staple of the genre – usually introduced to underscore human frailty and magnify the sense of threat. I don't believe in the supernatural, nor in premonitions, but life imitates art, and all the indications were that something bad was about to happen. I should have read the chapter headings. I should have seen where the story was going.

*

Rachel was a single parent with two children, Sabina, aged five, and Sean, aged eight months. She had separated from her Austrian husband and had returned to England in order to live near her parents, Bill and Ursula, who had retired and moved to the village two years earlier. Rachel lived with her younger sister, Sonia. Their brother, Warren – a late addition to the family – was eighteen, and lived with his parents.

My wife and I got to know Rachel and Sonia quite well. We all had a lot of time on our hands and were constantly paying each other social calls. We talked, watched the children playing on the floor, smoked cigarettes and drank tea.

It soon became evident that Rachel and Sonia were deeply dissatisfied. Rachel missed living in Austria. She had grown very fond of the Austrian way of life – skiing, pastries and coffee houses – and now she felt trapped and bored. When her marriage broke down she had had no other option but to return to England – to grey skies, chores and childcare.

Sonia's circumstances were different to her sister's but also unhappy. She had been having an affair with a married man called Henry for several years. Henry had promised to leave his wife when his children were older; however, he wasn't prepared to commit himself to a precise date. He owned a moderately successful haulage business and lived some 60 miles away in a coastal town. Occasionally he would appear in a smart convertible and take Sonia out for the day. Bill and Ursula didn't approve of this arrangement – they were devout Christians – but Sonia didn't care. She was in love.

Rachel and Sonia found it difficult to understand why my

wife and I had chosen to exchange life in London for the village. To them our decision was incomprehensible.

'Why on earth did you come to this godforsaken place?' Rachel asked, lighting a cigarette and blowing the smoke across the kitchen table.

'We wanted to get away from the city,' I replied.

'But nothing happens here,' she said.

'That's what's so appealing.'

She shook her head and scooped up her son. 'It drives me mad . . . '

'Don't you find it beautiful?'

'No.'

'When I lived in London, I used to look out of my kitchen window and see a brick wall – just a few feet away. It felt claustrophobic. Now when I look out of the window I can see *that*.' I pointed at the giant, hunched shoulder of the fell that overlooked the village. The summit was capped with fingers of ice that extended down the slope and touched a winding river of scree.

Rachel peered out of the window and took another drag of her cigarette. 'It's grim.' Blue-grey smoke coiled out of her mouth.

'Well . . . today maybe.' Rachel lifted her mug and took a sip of tea. I felt obliged to continue. 'I've always wanted to live somewhere where it's possible to see the seasons changing – to get in touch with something . . . real.'

'Wasn't London real?'

'Not in the same way.' I paused before making a final, positive assertion. 'I like this place.'

'Give it a year . . . '

A few days later I had a similar conversation while walking to the dairy to buy some milk with Sonia. The air smelled of manure and wood-burning stoves. It was raining hard and the road was like a quagmire. A farmer – who spoke in an incomprehensible dialect – herded his cows through the village every day and the tarmac was always covered in clods of earth. My boots squelched with each step.

'Look at all this shit,' said Sonia. 'You can't think this is good – surely.'

'I suppose the weather could be better.'

'Don't you miss civilisation?'

I found myself pointing at the fell again. 'Look . . . '

Sonia threw me a sideways glance, checking that she'd understood me correctly. Then she gazed up at the rounded mass that loomed above us. She blinked and wiped the rain out of her eyes. 'What about it?'

'It's been there for millions of years.'

'Of course it has. Where else would it be?'

'It gives me . . . I don't know – perspective. Don't you *feel* anything – when you look at it?'

'No,' she said, taking a certain amount of pleasure in her obstinacy. 'It's a fell.'

At night, the silence was absolute, and without light pollution it was possible to see meteors. Standing on the hill behind my cottage, I would watch streaks of luminescence dissolving in the sky. A full moon would transform the land into a dreamscape. Across the valley, a distant Victorian viaduct became an exquisite ornament crafted from silver and glass. Constellations burned with fierce clarity. It was curious. I had left London to find something 'real' but life was feeling more and more unreal. Perhaps my wife felt the same way too. If so, she didn't say. We would sit together for hours, looking into the flickering flames of the coal fire, without uttering a single word. Neither of us had the courage to ask the obvious question: where is this going?

The season changed. Lambs appeared and filled the air with timorous complaints. 'Look!' I hollered, lifting my son from his buggy and pointing. He studied their antics with sceptical indifference.

I had immersed myself in books about folklore and had become fascinated by the local myths and legends. Many were stories of doomed love, but the majority concerned supernatural occurrences: screaming skulls, witches turned to stone, visitations. I wrote a short talk for radio on this topic, sent it to the BBC, and a few weeks later an envelope landed on my doormat with a cheque for twenty pounds inside. This was the first time anyone had paid me for writing anything and I was deliriously happy.

One day, Sonia unexpectedly announced: 'Rachel's met a bloke.'

'What's his name?'

'Luke. He's from America.'

It seemed implausible under the circumstances.

'An American? Here? Where did she meet him?'

'Outside the Kings Arms.'

The Kings Arms was a hotel in a small market town situated about 12 miles west of the village. Rachel had been shopping on the high street when a man had approached her. He had started a conversation and they had talked for about thirty minutes. At the end of their conversation Rachel had invited him to dinner.

We didn't see Rachel for a while, but we received regular bulletins from Sonia.

'He's some sort of preacher.' Sonia's nephew – Sean – was sitting on her lap. He was dribbling and she used her handkerchief to wipe his mouth and chin. 'He's come over here with some other people – members of the same group – to organise meetings and stuff like that. Mum and Dad were interested, of course, and they've all started praying together.'

'I didn't think Rachel was very religious.'

Sonia raised her eyebrows. 'She isn't very religious. Well, not *that* religious.'

'What does Warren think?'

'He doesn't care. He just disappears with his mates.'

'What do you think?'

Sonia's eyes expressed her misgivings without need of speech: *Isn't it obvious?* She sighed and wiped her nephew's mouth again.

Over the next week or so I would often look out of the window and see Rachel walking through the village – on her way to her parents' house – arm in arm with a tall man. Sometimes, they would be accompanied by a small, casually dressed group, a slim woman with long sandy hair and two men. They seemed to be keeping a respectful distance, following Luke and Rachel, smiling benignly.

Henry arrived in his convertible and took Sonia away for a few days. She had had enough of having to do extra childcare and wanted a break.

My wife finally got her bar job. I drove her into the market town before opening time and thereafter spent most evenings staring into the fire on my own.

We were sitting at the kitchen table when there was a knock on the door. It was Rachel – and she'd brought Luke with her.

'Come in,' I said.

As Luke entered he had to lower his head to miss the lintel. He was in his early thirties and dressed in a blue check shirt, jeans and trainers. He was clean shaven but had hair that was beginning to cover his ears and creep over his collar.

Rachel and Luke sat down on our sofa and we offered them tea. They accepted and we made small talk for a while. Luke was from Georgia but he didn't speak with a slow southern drawl. In fact, his delivery was animated – excessively so – and accompanied by expansive gestures. Rachel didn't say very much and seemed quite content to let Luke dominate the conversation. There was something about her uncharacteristic reticence that reminded me of an awkward teenager. She giggled, stroked Luke's thigh and intermittently rested her head on his shoulder – emitting audible, amorous sighs. I noticed Luke's hands – his large knuckles – and the way he clenched his fists to emphasise particular points.

'So what brings you to this part of the world?' I asked.

'The Lord's work,' he replied.

'Yes, but why here?'

Luke leaned forward and spoke with manifest certainty:

'I opened my heart to Jesus and by the grace of His loving kindness, He gave me direction. He always does.'

What was Luke saying? That he had received instruction directly from God? And that God had told him to start a mission in a shabby, nondescript English market town?

Rachel detected my discomfort. She sat up, smiled and said, 'Listen, we've got some really exciting news.'

'Oh?'

Luke and Rachel looked into each other's eyes and then laughed. The intensity of the previous moment dissipated. 'We're going to get married. As soon as my divorce comes through – we're going to get married.'

'Congratulations,' I said, struggling to conceal my surprise.

Rachel took Luke's hand and squeezed it. Their manic grins were expanding.

'I'm sure you'll be very happy together,' said my wife. She was equally unconvinced. I could hear the strain in her voice.

'And when Luke's finished his mission here,' Rachel continued, 'we're going to live in America. Luke's parents have a ranch. Can you imagine how great that'll be for the kids?'

'I feel blessed,' said Luke. 'Truly blessed.' His fingers came together and I could see that his instinct was to lead us in a communal prayer of thanksgiving. But he checked himself and said, very simply, 'I'm a lucky man.'

Sonia returned the next day, and came to see me the following evening.

As soon as I'd opened the door she said: 'Have you heard?'

'Yes,' I replied.

'It's mad, isn't it?'

'Have you spoken to your parents?'

'They're a bit worried. But they're believers . . . and the Lord works in mysterious ways – doesn't He? Christ – and I thought I was bored.' She considered her sister's sudden conversion to evangelism dubious.

'Rachel must feel something for him . . .'

'She says it was love at first sight. She says she feels like a new woman. But she's done this before. She persuades herself that she's in love and off she goes.' Sonia traced the ascent of a plane in the space between us. 'It's all very convenient.'

'Have you told her what you think?'

'Yes.'

'And how did she respond?'

'She said: What about *you* and Henry?' Sonia forced an acid smile. 'I met Henry three years ago. Rachel met Luke five weeks ago. If I *am* making an idiot of myself – and I accept that's a possibility – then at least I'm making an idiot of myself over a man I know.' She tapped her cigarette against the ash tray and took another drag. The tip glowed orange before she blew a thin stream of smoke through an angry pout.

'Have you spent much time with Luke?' I asked.

'Not a lot. When he comes over I make myself scarce. I take the kids out or go upstairs and smoke.'

'He's very . . .'

Her eyebrows climbed a fraction. 'Odd?'

I didn't want to judge him. 'He's entitled to his beliefs.'

'But why would God want him to start a mission here? Why not somewhere sensible like Africa?'

'Because God works in mysterious ways, I suppose.'

After expressing more concerns about her sister's prospective marriage, Sonia reflected on the weekend she'd just spent with Henry. He'd taken her to a hotel – a former stately home – with grounds and a spa. She'd had a good time, but being dumped back in the village had made her feel exploited and desolate. I tore a tissue from a kitchen roll and handed it to her before the first tears appeared.

'Thanks.' She dabbed her eyes. 'You should have been a therapist.'

As Sonia was leaving, she delayed on the doorstep and asked me how my wife was.

'She's fine.'

'Is the job working out?'

'She likes it – I think.'

Sonia glanced down at her wristwatch. 'What time does she get back?'

'Late.'

She nodded. 'Thanks for listening.' She surveyed the cottages with their black, lightless windows – sighed – and then marched off briskly into the night.

Summer arrived. From the village, it was possible to walk along a track that followed a hidden valley. I used to walk for miles without seeing anyone – past a Neolithic hill fort, over slopes littered with flint and bones – all the way to a bridge made from black and red stone that was so ancient, most of it had fallen into the river.

The sense of disconnection that I'd had since leaving

London was intensifying and was now accompanied by fleeting intimations of unease. I wondered how long it was possible to go on living like this. Surely something would happen – surely reality would catch up with me and demand that I participate in the world?

I was in the market town with my wife, pushing my son along the pavement in his buggy, when we happened to meet Luke. He was with three of his fellow evangelists. He introduced the sandy-haired woman who I'd seen in the village as Amber, and the two young men who I also vaguely recognised. 'Joshua and Nate,' said Luke. They were all American. After these introductions, Amber, Joshua and Nate took a step back, and my wife and I continued exchanging polite and superficial pleasantries with Luke. It was curious how the other three had retreated, only to stand together in silence, their faces rigid with identical fixed smiles.

We made our excuses and continued walking. When we were sufficiently distant not to be overheard, my wife said. 'Those people – they're like disciples.'

'Yes,' I agreed. 'They are.'

After laying my son in his cot I waited in the shadows for the rhythm of his breathing to change. When I was satisfied that he was asleep, I descended the stairs and sat in the lounge with my wife. She was reading. The resolute silence was broken only by the rustling of pages. I turned on the radio, motivated perhaps by some marginally acknowledged need to distract myself from discomfiting thoughts. The radio reception was poor, and the piano music – it may have been a Chopin nocturne – vied with waves of interference and ghostly foreign voices. I fiddled with the dial to get a better signal but it was hopeless.

Suddenly, there was a frantic hammering on the front door. It was so loud that my wife and I started. The hammering continued; short, irregular bursts. We never had late visitors.

My wife said: 'Who can that be?'

I pointed at the ceiling, tutting. 'They'll wake him up.'

I leapt from my seat and into the narrow hallway. 'All right,' I called, as I turned the key. The bolt clicked and I pulled the handle.

Sonia was standing outside. Her eyes were bright with terror and she was breathless. 'Please help,' she said. Her make-up was smudged and her lips were trembling. 'Please help.' She was so frightened she could barely speak.

'What's happened?' I asked.

'It's Luke.' Her voice had acquired a whining quality, like a child about to cry. 'He's going to kill us. He wants to sacrifice us to God. Please, you've got to help.'

I looked at my wife. 'Lock the door.' She nodded and I waited for the sound of the bolt. Then I checked to make sure the door was secure. 'Okay.'

Sonia set off, keeping her head low and glancing nervously over her shoulder. I followed a few paces behind. 'Where is he now?'

'I don't know. He tried to break the door down – he's gone completely mad.'

We kept to a pathway that ran parallel to the road because it offered some concealment. I could taste fear in my mouth – iron in my saliva. I can remember thinking: *This can't be happening. This sort of thing doesn't happen in real life.* The world remained stubbornly solid. I kept moving, motivated not by courage but by social anxiety. If I turned back and my cowardice resulted in the death of two women, a child and a baby, then it would be (to coin a typically English phrase) frightfully embarrassing.

I was far from happy about having left my wife and son to fend for themselves. What if Luke changed his mind and decided to sacrifice them instead? It occurred to me that turning back would not be entirely indefensible; however, the likelihood of Luke returning to Rachel's house was greater, so with considerable reluctance, I kept putting one foot in front of the other.

The windows of the cottages that lined the roadside were dark. These buildings were mostly occupied by retired couples and members of the farming community, all of whom tended to go to bed early. I still expected to see a few strips of light between drawn curtains but the village looked as if it had been abandoned.

We reached the end of the path and Sonia hesitated before advancing into the open. She peered around the edge of a wall and immediately withdrew again. 'He's there,' she whispered. 'Shit. He's there.' She began to sob and covered her mouth to stifle her whimpering. We changed places, and when I looked I found it hard to believe what I was seeing. I had always enjoyed horror films, and now, apparently, I was in one. The scene was perfectly composed: a generic cliché. Once again, I thought: *This can't be happening.* It was almost parody.

A mist was rolling off the fell and dispersing through the village. At the end of the road was a bulb suspended between two posts. It was producing a hazy light which enveloped a tall figure. Luke's head was thrown back and he appeared to be talking to the sky. The horizontal continuity of his arms suggested identification with Christ on the cross. Slowly, he reached upwards, his fingers becoming claw-like. Then, he came forward, adopting the stilted gait of a B-movie monster.

'We've got to go,' I said to Sonia, grabbing her hand. 'We can't stay here.'

The road was exposed but it was also very dark. Luke didn't seem to notice us as we made our escape and when I glanced back I saw that his lumbering progress was slow. His arms were still raised and he looked particularly ghoulish as he emerged from his glowing shroud of mist.

When we arrived at Rachel's house, Sonia rang the bell. All of the glass panels in the door had been broken. Some shards – with red tips and edges – remained embedded in the woodwork. Scraps of skin and lumps of tissue were hanging from the sharpest points. The front step was splashed with blood and rust-coloured smears covered the door frame. I felt slightly sick.

Sonia peered into the receding gloom. The streetlight was no longer visible but it had created a faint aurora. She pressed the bell again. 'Come on, come on.' Then she shouted through the broken glass. 'Rachel – it's me – open the door.'

Rachel sprang out into the hallway and rushed towards us. She produced a key, unlocked the door, and when we were inside, secured the door behind us.

I discovered Sabina in the lounge. The child was standing very still. Her pupils were so dilated her irises had become two black, shiny circles. She didn't respond when I greeted her. Sean was grizzling, propped up on a pile of cushions.

Rachel put her arm around Sabina and pulled her close before addressing me. 'Thank you. I'm really sorry. We didn't know what to do.' She took a deep breath and tried to explain. 'It was so frightening – I've never been so scared in my life.'

'That's *his* blood – on the door?'

Rachel nodded. 'He'd been here for a couple of hours and we were just – you know – talking – like we normally do – but he wasn't right. He wasn't making any sense and he kept on stopping to say prayers. And then he said that maybe we didn't have to wait – to be married, to be together – that there was another way . . . and that Sabina and Sean could be with us too . . . in heaven.' Rachel stroked Sabina's hair and began to cry. 'I got really scared and asked him to leave, but he wouldn't go. He started to get angry – told me that I shouldn't have any doubts – that it was wrong to doubt – and that I should be strong and trust him. I said I needed a few minutes on my own to think it over and asked him to go outside. When he heard me locking the door he went crazy. It was horrible. He tried to break the door down. He just tried to punch his way through.'

Luke had eventually given up and walked away, presumably in order to commune with God and receive instruction.

'Where's Warren?'

'With my dad – they've gone out tonight.'

They were attending a big social event being held in the

next village. This was another reason why so many of the cottages were dark.

I wasn't sure what I was expected to do. If Luke succeeded in breaking the door down I might be able to buy Rachel and Sonia a little time. But it would probably be only a few seconds, particularly if God had advised Luke to find himself an axe.

'Oh God,' said Rachel. 'What have I done?' She threw a guilty glance at her baby son.

My mouth was dry and my legs were shaking. I felt hopelessly inadequate and I was so anxious by this time that my head had emptied of thoughts and I was experiencing a kind of numb detachment. It was as though my mind was shutting down.

And then, suddenly, everyone was screaming. Sean started to wail too.

Rachel, Sonia and Sabina were all looking in the same direction. A pale oval was hovering in the darkness framed by the window. I could taste fear again – a mineral toxicity that caught at the back of my throat. Features clarified as the face pressed up against the glass. I heard Rachel exclaiming: 'No – it's all right – it's all right – it's Warren.'

Sonia laid a hand on her chest and said, 'I can't take much more of this.'

Rachel signed to her brother and went to open the door. I heard Bill's voice: 'Jesus – what's all this?'

'Did you see Luke?' Rachel asked.

'We passed him in the car . . . '

Bill and Warren entered the room and I felt hugely relieved. Everyone was talking but I wasn't really listening.

The situation was no longer my responsibility. I just wanted to go home.

Outside, a car was parked in the road and a few young men were inspecting the blood and broken glass. I guessed they were Warren's mates. I had taken only a few steps before I came to an abrupt halt. Luke was standing at the garden gate. I sensed a general withdrawal – people stepping back – as Luke came forward. We met each other halfway along the garden path.

'Hello Luke.'

He looked down at me. His expression showed a hint of recognition but he seemed distracted. His head rocked back and he gazed upwards, his eyes locking on the highest stars. He started whispering very fast. All that I could hear at first was sibilance, but as I listened, his accelerated speech became interpretable. 'Father, Father, Father – Your house has many rooms, Your room has many houses. Did You not tell us so? Father, receive us into Your love. For the day is coming – give us this day . . . this day of days . . . For Thine is the Kingdom, the power and the glory. Deliver us from evil. Surely – surely – He has borne our grief – and carried away our sorrows.'

His shirt sleeves were in tatters, black with blood, and deep cuts covered his forearms. Something, I couldn't see what it was exactly – a shred of muscle or an artery – was actually protruding out of a long, open wound.

'Purify us from all unrighteousness. And I shall be Your faithful witness – now and forever – Amen. Thank you, Father – thank you, thank you.'

'Luke,' I said, 'perhaps you should sit down. You've lost a lot of blood.'

He rotated his arms in the light spilling out from the hallway. His hands were completely red. 'By these marks,' he said with solemn confidence, 'you shall know me.'

'All the same, you really should sit down.'

I was surprised by his reaction. He fell to his knees.

'Maybe you should keep your arms high too,' I suggested. 'You're still bleeding.' Again, he did exactly as I'd asked. 'How are you feeling, Luke?'

'Good,' he said. 'Good. Delight yourself in the Lord and He will give you the desires of your heart.'

For a few seconds, his teeth chattered.

'Are you cold?'

'No . . . I'm not cold.'

I hoped that someone was calling the police while I kept him occupied.

Luke continued muttering Biblical quotations and fragments of prayer. Occasionally, however, he would fall silent and give me a curious, probing look. It made me uncomfortable. But if I asked him a solicitous question, the words would soon start flowing again, and he'd quickly revert to addressing the sky. As another episode of chaotic prayer petered out, his head lolled sideways and his eyes became inquisitive. Before I could give him a prompt he said, 'So, tell me something.' His voice sounded quite normal and conversational. 'Because I've been wondering . . . is your wife an obedient woman?'

'Obedience isn't something we talk about.'

'Isn't it?'

'No. It isn't something I'd expect.'

'How come?' I shrugged. 'Is that the truth?' His voice – now a preacher's voice – was sincere and sympathetic. 'Honestly? It doesn't matter? It doesn't matter that *you* don't wear the pants in *your* house? A thing like that! It doesn't matter?'

'I guess not.'

He considered my answer and nodded. After a few seconds he said, very quietly, 'I know who you are.'

I bent down to hear him better. 'I'm sorry?'

Our faces were close. I was aware that he had started to smile. The corners of his mouth had curled upwards but his eyes remained narrowed and suspicious. He lunged at me and shouted: 'Fuck you – Satan!'

I jumped back, shocked. His hands raked the air. He made a few more efforts to grab me and then the fight seemed to go out of him. He sat back on his heels and bowed his head. 'Father,' he whispered. 'Thank you.'

The mist pulsed with blue light. A police car arrived and two officers leapt out. I had briefly heard the crackle of their radio. 'I'm just a neighbour,' I said, pointing at the house. 'The family are inside.' I gave Luke a wide berth and hurried down the road.

When I got to my cottage I paused to survey the scene. The light suspended between the poles flickered. I thought of the hill behind me, its dark invisible mass, and the earth – saturated with ancient, sacrificial blood.

I heard my wife say: 'Who is it?'

'It's me.'

She opened the door and I stepped into the kitchen, where I immediately collapsed on a chair. I was emotionally and physically exhausted.

'What happened?' my wife asked.

'Can you make me a cup of tea?' I asked, hoping, on this occasion, for obedience.

Catastrophic breakdowns are the result of stressful life events interacting with vulnerabilities, which may vary from person to person and can be psychological, biological, or both. It seems highly likely that Luke was predisposed to psychosis. That he had chosen to bring his disciples to an obscure English market town raises serious questions about his prior mental health. This sleepy backwater was no Gomorrah and the ordinary people who lived there had no obvious need for spiritual salvation.

Luke's mission was redundant and oddly random. His Messianic affectations were grandiose and although he only spoke of Jesus giving him guidance, the clear implication was that some kind of dialogue was taking place long before he left America. Auditory hallucinations do not necessarily predict the development of serious psychiatric illness. Some people are able to correctly identify them as internally generated phenomena and live regular lives; however, when voices are attributed to God, the context is almost always delusional.

There is a considerable overlap between religious experience and psychosis. If you are a devout Christian and one day Jesus tells you to travel to another country and do his bidding, why wouldn't you? And if he also told you to kill in his name, why wouldn't you do that too? The Bible is full of Godly violence. How is it possible to discriminate between the authentic voice of God and an auditory hallucination? Ultimately, if you are a believer, you probably can't. On the other hand, if – as Freud suggested – all of religion can be understood as an infantile defence against the harsh realities of existence, there is no dilemma. When God speaks to you, it is always an auditory hallucination, because He doesn't exist.

I have always exercised caution when talking to patients with deeply held religious beliefs – particularly if they have a different heritage. What passes for normal in one culture can be mistakenly classed as abnormal in another. I was once working at a drop-in centre for a mental health charity. I was asked to assess a middle-aged Indian woman whose English wasn't very good. She seemed to be suggesting that she could hear the voices of Hindu deities: Lord Shiva, Hanuman the monkey god. I spent several hours with her, being careful not to ask leading questions and doing my best to clarify the nature of her experience. I was mindful of my Western, secular biases, and at the end of the interview I was undecided. Consequently I spoke with her husband – who was also Indian and Hindu. I explained how I didn't want to make any mistakes because of cultural differences. 'Isn't it obvious?' he said impatiently. 'She's bloody mad.'

Falling in love can destabilise individuals who exhibit none of the risk factors associated with psychiatric illness. Even the most grounded, well-balanced and rational people become lovesick. Luke – who was already prone to delusional thinking and hallucination – was simply unable to cope with the intensity of love. Falling in love was the life event that interacted with his vulnerabilities and caused his breakdown.

Luke, being an evangelical Christian, could not countenance sex before marriage and he was morally obliged to postpone intercourse indefinitely. Rachel had initiated divorce proceedings, but there was a strong possibility that her husband would be uncooperative and cause delays. I doubt that Luke had had many experiences of falling in love. He was unprepared for the upheaval, the longing and the yearning – the sleepless nights. And most of all, he was unprepared for the onset of desire – restive urges – lust.

Wilhelm Reich, perhaps the most colourful figure in the history of psychiatry, believed that mental illnesses are caused by various forms of sexual frustration. Sexual energies can be blocked or find insufficient release. Orgasms can be unsatisfactory. This viewpoint has much in common with Freud's early proposal that a build-up of libido in the body causes anxiety. Freud assumed an underlying biological mechanism that resembled the conversion of wine into vinegar. In the decades that followed, his understanding of mental illness became more refined and he abandoned this position. Reich, however, remained faithful to Freud's original hypothesis and became convinced that sexual blockages could also cause physical problems. When he was visited by an elderly woman who suffered from a diaphragmatic tic, he taught her to masturbate and the tic disappeared.

Reich was a very modern thinker in that the approach he advocated was holistic. He recognised, for example, that psychological defences are sometimes embodied. When we repress, we become tense. It is as though our muscles harden to become a kind of armour. This observation led him to develop innovative treatments involving massage.

He discovered that inhibitions and blockages could be released by manipulating the body. These innovations were not welcomed by the majority of psychoanalysts, for whom touching patients was strictly taboo.

Being Jewish, Reich chose to leave Europe for America in 1939 to escape Nazi persecution.

The interventions that Reich developed – collectively and unappealingly named 'vegetotherapy' – have not gained widespread acceptance. This is largely because his ideas became increasingly outlandish and he lost all scientific credibility. He reconceptualised libido as a cosmic life-force and claimed that it could be collected in 'accumulators' and used to cure cancer. He constructed massive energy-cannons that he aimed at the sky in order to produce rain. He also used them to destroy UFOs and protect the earth from alien invasion. In the final years of his life he attracted the attention of the US authorities. His accumulators were destroyed, his books and journals burned, and in 1957 he died in jail.

I don't believe – as Reich would have – that sexual frustration provides a full explanation for Luke's breakdown; however, I think it played an important part. He was at war with himself. His rigid beliefs about the impropriety of pre-marital sex were preventing the satisfaction of his instinctual needs. Frustration was intolerable, but so was the alternative – sin. This kind of dilemma, which requires an individual to make a choice between two punishing outcomes, is known as a 'double bind'. Throughout the 1960s and 1970s, many psychiatrists and psychotherapists were convinced that double binds (usually created by dysfunctional communication within families) were a cause of schizophrenia. What was Luke to do? The solution to his predicament was transcendence. He would become one with Rachel in the kingdom of heaven. And I suspect that the voice in Luke's head – the voice of God – was approving. Sex is a pale imitation of that great romantic ideal – the union of souls – a state that can be accomplished only by escaping corporeal limitations.

If Rachel had failed to get Luke out of the house, things might have turned out very differently. He might have locked the door, removed a knife from a kitchen drawer, and killed Rachel, Sabina and Sean. He might also have killed Sonia. And after sending them all to heaven, he would, of course, have followed. Love has unforeseen consequences. We should always take it seriously.

Luke was hospitalised and I never saw him again. His parents were contacted and, when Luke was fit to travel, they took him home. His disciples – bereft of purpose and direction – also returned to America.

Ever since my arrival in the village, I had been disturbed by a gnawing sense of unease. I had attributed it to my fascination with the local folklore. I had been reading far too much about haunted ruins and travellers seized by demons. I had been spending far too much time alone. Now, I felt relieved. The boil had been lanced – the poison drawn. The bad thing had happened.

But I was wrong. The bad thing hadn't happened.

A few days passed. Then my wife told me that she wanted a divorce.

Freud once famously posed the question, 'What does

woman want?' He was dissatisfied with his understanding of female psychology. For thirty years he had studied the 'feminine soul' and he was baffled. This quote appears frequently in books – books mostly written by men. I think we find it consoling. If Freud didn't know what women want, what chance the rest of us? Although in reality, of course, his ignorance exonerates no one.

Life does not progress in an orderly fashion. Periods of stability are interrupted by critical events that precipitate change. Joseph Campbell – who wrote extensively on the subject of comparative mythology – pointed out that most stories reach a point where a crisis or blunder alters the direction of the protagonist's life. The critical event, the turning point, often involves entering a dark forest and encountering a figure – sometimes magical, sometimes sinister and dangerous – who is effectively the herald of change: 'The familiar life horizon has been outgrown; the old concepts, ideals, and emotional patterns no longer fit; the time for the passing of a threshold is at hand.' The crisis, according to Campbell, is also a 'call to adventure'.

There is tremendous wisdom to be found in the symbolism of folklore and myth. Indeed, one could argue that many of the discoveries of psychotherapy are not discoveries at all, but the transliteration of helpful principles encoded in stories: I had entered the dark forest of a failing marriage – lost my way – and encountered the herald of change in the form of a psychotic American evangelist. At the time, it seemed to me that my life had completely unravelled. I was unhappy, uncertain and ill-prepared for the emotional demands and legal wrangling that lay ahead. I wasn't in great shape. Luke was still lumbering through my dreams, emerging from wreaths of mist, drenched in blood – a sibilant malevolence.

But crises are catalysts. They move us on – break us up – so we can be recast in new forms. The herald of change does not appear at *any* point in a story. He or she appears when the old 'emotional patterns no longer fit'. If I had been familiar with Joseph Campbell back then, I might have found solace – rather than supernatural dread – in all of the folktales I'd been reading.

Two months later, I was sitting in a lecture theatre, pen in hand, a note pad balanced on my knees, in a very ready state to learn about the human mind and relationships.

Chapter 7

The Stocking Game: *Dr B and Fräulein O – a cautionary tale*

Cassandra always dressed in the same combination of clothes: T-shirt, tight black jeans and trainers. She rarely wore make-up, and when she did it was only just noticeable – hints of pastel around her eyes, a subtle gloss on her thin lips. She moved with an easy, loose-limbed grace, and would often raise one heel onto the edge of her seat and hug her knee. Her hair, a shade of brown, was long and straight and always flopping in front of her eyes. She parted her fringe with an equine shake of the head. When she entered the consulting room, she would narrow her shoulders and allow her fawn-coloured raincoat to drop to the floor. Its descent was usually assisted by a weighty paperback curled into one of the pockets.

At the age of sixteen, Cassandra had developed an eating

disorder. This went into remission after a year; however, from that time onwards, she had experienced intermittent episodes of mild depression. She was now in her twenties and I was seeing her for a short course of structured psychotherapy. Every fortnight, I reviewed her mood and thought diaries, checked her activity schedules and set new goals. Cassandra liked working in this systematic (if somewhat inflexible) fashion. She had received exploratory, non-directive counselling when she was at school and had found the approach woolly and unhelpful.

It was a hot summer's day and I had opened the window. A breeze provided welcome relief from the stifling heat but the traffic was extremely noisy. Frustrated drivers – unable to make progress – were revving engines and sounding their horns.

'I've had something on my mind for a while,' Cassandra said, momentarily distracted by the billowing net curtain. 'I haven't written about it in my diaries.' I had been studying a sheet of paper on which she had kept a record of her thoughts and feelings. I looked up and she continued. 'I didn't think it was appropriate – or relevant, even. But it's becoming an issue. I need to talk about it.' She raised one leg and hugged her knee. 'I've been seeing someone. It's been a while – a couple of months now, at least.'

'Where did you meet?'

'In a park . . . I was jogging and I'd stopped, just to catch my breath, and there was this guy – sitting under a tree – strumming a guitar. He said hello and we got chatting. He was from Australia and he'd travelled quite a lot: South America, China, Bhutan. I've been meaning to go travelling.

It's something I've wanted to do for ages but I never seem to have enough money.' A car horn, sustained for several seconds, was joined by two more. 'He was interesting,' Cassandra persevered, 'and a lot of fun. He started singing songs – his own material – and he was really good.' She was oddly insistent about this point. 'I mean *really* good. He used to gig in Sydney.' She swallowed and added: 'His name's Dan.'

After Dan's impromptu concert they had gone for a coffee. Dan invited Cassandra back to his flat and later the same day they had made love.

'He's really easy to be with. I haven't met anyone like him for ages – not since university. We stay up late, talking about music, philosophy, art.' She made her voice comically ponderous. 'The meaning of life . . . it gets really intense sometimes, but I think that's a good thing. He's reminded me of how much I used to enjoy that kind of conversation. My friends just aren't interested any more. All they want to talk about is shopping and money and careers.' She released her knee and sat normally. 'He introduced me to one of his friends – Emily – who was nice. She's from Australia too. Anyway, he was saying how sad it is that we all end up living dull, conventional lives – how we get brainwashed – and how we'd all be a lot happier if we were more open to experience, willing to experiment with different ways of living. And then he asked me how I felt about . . . ' She stretched the hem of her T-shirt. 'Well . . . having a threesome with Emily. And I thought, yeah – why not? I mean, he was right. We do just shut ourselves off from new experiences; we are all brainwashed. So . . . he spoke to Emily and set it up. Emily came round. Dan rolled some joints – I didn't smoke any – weed messes with my head too much – and the three of us went to bed. It happened again last week.' Cassandra tapped her finger tips together. 'I'm not stupid. I knew what was likely to happen. But Emily – she comes on too strong – and on both occasions, Dan just backed off and watched. And – the thing is – I'm not really bisexual. I go along with it, but it's not really me.'

'Have you told him that?'

'Yes. And he said fine – that's okay. But I know that he wants us to carry on.'

'Perhaps you should talk to Emily.'

Cassandra didn't acknowledge my suggestion; her attitude became introspective. When she spoke, it felt like she was finishing a syllogism that had been worked through in silence. 'I *really* like him.' Her voice betrayed her: a lurch of emotion – capsize, floundering, drowning.

'Are you worried that he'll see you differently now?' I asked. 'That he'll lose interest – and the relationship will end?'

Her assent was as subtle as a vibration. The tension went out of her body and she allowed herself to fall back into the chair. Her arms were hanging by her sides, palms facing outwards. There was something almost post-coital about her abandon. She looked at me down the length of her nose with provocative hauteur: 'I love the way he does it, the way he moves. He never rushes.' Her eyes closed and opened. 'He never rushes.' The repetition was devotional – like an incantation. The car horns reproduced a brassy, discordant syncopation. 'I don't want it to end. Not now. Not yet.'

*

Sylvie was in her early thirties and had been troubled for several years by a pervasive sense of dissatisfaction. She didn't know what she wanted from life any more. She felt directionless, unmotivated. Her discontent was slowly fermenting into depression. 'I feel trapped,' she said, making a cage with her fingers and inviting me to imagine an imprisoned bird inside. 'I don't know how this happened. I wasn't like this when I was younger.' Sylvie often made unfavourable comparisons of this kind, contrasting present and past selves. 'I was so much more alive then.'

When she was eighteen, Sylvie had worked as an au pair for a wealthy couple called Peter and Amy. The couple had two small children. A condition of employment was travel to Greece, where the family owned a villa on one of the islands. They spent every summer there and Sylvie was delighted to go with them.

'It was fabulous. I loved it. Every day, Peter would take me out in the yacht. He'd moor in some cove and we'd jump off the side and swim. We were spending quite a lot of time together – alone – and . . . I suppose he seduced me. I didn't put up much resistance. Actually, I really fancied him. It became a regular fixture and I began to feel quite guilty. I didn't like deceiving Amy. She'd always been good to me. We got on well together. I told Peter that I was unhappy with the situation and he said that I shouldn't worry – that Amy wouldn't mind. They had an understanding. A few days later, Amy took me aside and said she knew what was going on and – as far as she was concerned – it wasn't a problem. I guess it felt a bit weird having that conversation, but – to tell the truth – everything felt weird.

I'd left my normal life behind and it felt like I was living in a dream.'

The blazing sun, the scintillating blue of the Aegean Sea, the sky a psychedelic shade of purple: the yacht at anchor, the burning sand, a young woman – her nudity translated into a kind of abstract brilliance – and an older man.

'Peter started coming to my room. He'd knock on the door and I'd let him in. He never stayed until morning. He always went back to Amy. This carried on for a week or so. Then, one night, Amy joined us. They never asked me how I felt about this – there was no discussion. And they'd obviously planned it. Peter wasn't at all surprised when she appeared. I should have felt manipulated – exploited. But I didn't. It was great – really great. And I felt so alive.' She stroked her collar bone and smiled, coyly. 'There was so much . . . contact.' She crossed her legs – allowed the heel of her shoe to detach itself from her foot – and waited for my reaction.

Freud advised psychotherapists to model themselves on surgeons, to lay aside any feelings that might disrupt concentration or hinder technique. But that's easier said than done. Listening to Cassandra and Sylvie, I can't say I was unaffected by their confessions. Their narratives suggested pictures that formed in my mind. I wasn't emotionally inert.

Eric Berne – the psychiatrist who developed transactional analysis – identified a number of 'games people play' in social situations. These games – or entrenched patterns of behaviour – can appear innocent but often serve an ulterior motive. In 'The Stocking Game', a woman raises her leg in the presence of others and remarks: 'Oh my, I have a run in my stocking.' This manoeuvre is calculated primarily to attract attention and cause sexual excitement. Berne's writing – in this context at least – feels a little regressive, a reversion to a time when men were uneasy and mistrustful of female sexuality. However, human beings – both women and men – frequently employ such tactics (either consciously or unconsciously) to enhance self-esteem, exercise power, or control others.

A female colleague told me about a patient of hers, an athletic, muscular man who had entered therapy because he no longer found his wife attractive. He frequently made references to the size of his penis. In order to demonstrate that his boast was not an idle one, he started attending sessions wearing clingy sportswear. His favoured position was low in the chair with his legs spread wide apart.

The standard cartoon image of psychoanalysis, which shows the analyst, sitting out of view, listening intently to a patient lying on a couch, owes its existence to various late-nineteenth-century permutations of the stocking game. Freud tried placing his chair in various positions before finally deciding that the safest place to sit was next to the head of the couch. Some of his female patients were behaving provocatively and he wanted to keep well out of harm's way. His friend and mentor, Josef Breuer, had once underestimated the power and significance of patient sexuality and had consequently paid a high personal price. Freud didn't want to make the same mistake.

Psychoanalysis begins not with Freud, but with Breuer. This is a contentious statement, because firstly, Breuer never really practised as a fully fledged psychoanalyst, and secondly, many psychiatrists and neurologists were conducting similar therapeutic experiments before Breuer; however, Breuer's treatment of the young woman known as 'Anna O.' influenced Freud's thinking profoundly and the case study documenting her treatment, which Breuer wrote up many years later, set a stylistic precedent.

Breuer was a successful general practitioner and medical researcher. It was while he was working in the laboratory of the famous physiologist Ernst Brücke that he first became acquainted with Freud (his junior by fourteen years). On 18 November 1882 Breuer told Freud about his treatment of Bertha Pappenheim (a friend of Freud's fiancée and destined to be immortalised as Anna O.). Over a period of approximately eighteen months, Bertha had displayed a quite sensational range of hysterical symptoms and behaviours. These included headaches, loss of hearing, coughing, squinting, impaired vision, paralysis, cramps, cleaning rituals, tubercular emaciation (anorexia), hydrophobia, stiff joints, muteness, mood swings, agitation, violence, confusion, stupor, speaking only in a foreign language and attempted suicide. Sometimes she would slip backwards in time, avoiding or bumping into furniture that had been moved. And she would lapse into dream-like states and experience terrifying hallucinations. No physical cause for any of these symptoms was discovered.

Every evening, Bertha entered a trance-like state and muttered incomprehensibly. Breuer discovered that if he said certain phrases – repetitions of things she had said earlier in the day, or words charged with special meaning – her language became increasingly coherent until it became clear she was telling a story. The stories she recited reminded Breuer of those written by Hans Christian Andersen. When Bertha finished her stories she became calm, cheerful and lucid. She described this procedure as the 'talking cure' – a term we now use to describe all forms of psychotherapy. The benefit, however, was short-lived, and her condition deteriorated over a period of several days, until she was once again hallucinating and muttering in a trance.

Breuer's great therapeutic breakthrough involved hypnosis and the retrieval of memories. He was able to establish that each of her symptoms was related to a forgotten traumatic event that had occurred while she had been nursing her dying father. Her blurred vision and squinting, for example, were attributable to memories of crying. Sometimes, Bertha would actually re-enact her traumatic experiences – releasing her emotions by means of a cathartic theatrical performance.

Several years after Breuer had told Freud about Bertha, Freud was in a position to conduct his own therapeutic experiments. In due course, the two men collaborated on a major work titled *Studies on Hysteria*, which was published in 1895. The most important case in the book is that of Anna O. – the name given to Bertha to conceal her true identity. Feminist writers have made much of this moniker. It has been suggested that the palindromic 'Anna' represents the divided female psyche and that the O represents the madness of Ophelia or the ancient symbol for the female genitalia. The truth is probably less exciting: A and O are Bertha Pappenheim's initials (B and P) moved back a space in the alphabet. But there is more to decode than Pappenheim's clinical alias. Close to the end of the case study, Breuer makes an oblique admission: almost in passing, he says that he has 'suppressed a large number of quite interesting details'.

What details?

Bertha was twenty-one years old, pretty, petite – only 4 foot 11 inches tall – and endowed with the striking combination of dark hair and blue eyes. She was also extremely intelligent. She spoke five languages, painted pictures, wrote well, played the piano and had a passion for Shakespeare. For Breuer, going back home to Frau Breuer must have become increasingly dull, particularly after listening to Bertha's stories and enjoying her histrionics. Even though he lived only a short distance from the Pappenheim residence, he made a conspicuously large number of house calls. He saw Bertha every day for eighteen months and, needless to say, they became very close. For long periods, they would speak to each other in English – excluding others. She insisted that she be allowed to touch him – which he permitted. And when Bertha's father died, it was Breuer who soothed her agitation and put her quietly to bed.

Breuer's treatment of Bertha was officially concluded in June 1882. What happened immediately after is believed by many to be one of the 'quite interesting details' that Breuer chose to suppress. The reliability of documentary sources has been repeatedly challenged by historians of psychotherapy; however, these sources include Freud's recollections (as recorded by his biographer Ernest Jones) and Freud's personal correspondence. Breuer – we are told – was called back to the Pappenheim residence where he found Bertha going through the melodramatic agonies of a hysterical childbirth. In a letter written by Freud to the author Stefan Zweig, Freud says that Bertha cried out: 'Now Dr B's child is coming.' This was too much for Breuer, the respectable, trusted general practitioner with a reputation to consider. He hypnotised Bertha, calmed her down and fled, according to Jones, 'in a cold sweat'. He then took his wife to Venice for a second honeymoon; she had become jealous of Bertha and needed attention. Bertha continued to have problems and Breuer referred her to a colleague. He hoped that his 'patient, who has always meant a great deal to me, will soon be safely in your care'.

Freud was convinced that sexual feelings played a critical role in the development of hysterical symptoms. Breuer, given what had happened, wasn't overly keen to pursue this line of inquiry. Freud became frustrated with Breuer, their friendship suffered, and their collaboration came to an end. Many years later, Freud characterised Breuer's abandonment of Bertha Pappenheim as a form of scientific cowardice: he had retreated when he should have been pushing forward. As Freud slowly assembled the theoretical edifice that was to become psychoanalysis, he gave great significance to the sexual feelings a patient might feel for a therapist. He believed that such feelings should be discussed and interpreted because they were, in fact, displacements from childhood and formerly associated with the opposite-sex parent. Freud called the phenomenon 'transference' and came to regard it as an essential part of the therapeutic process. The reverse can also happen. When a therapist experiences sexual feelings for a patient it is termed 'counter-transference'. This is a problematic development and has no therapeutic value.

The notion of transference has broadened since Freud's time. All feelings (rage, anger, suspicion) displaced from any prior relationship can be usefully discussed in therapy. Analysing transference effects is a way of bringing historical problems into the present – the here-and-now – so that they can be dealt with more easily.

Breuer returned from his second honeymoon and resolved never to become so deeply involved with a patient again. He was a modest, unambitious man and, unlike Freud, was not overly concerned with posterity, which is perhaps why he allowed his young protégé to develop his ideas. Breuer's generous patronage set Freud on course for greatness.

Bertha continued to suffer from hysterical symptoms, but her subsequent achievements suggest that ultimately her health was restored. She published a book of children's stories, wrote a play and became a social worker, a reformer and what we would now probably describe as a 'feminist activist'. She translated Mary Wollstonecraft's *Vindication of the Rights of Woman* and was a founder member of the German League of Jewish Women. She travelled to Russia, Poland and Romania in order to rescue children whose parents had been murdered in anti-Semitic pogroms.

In late-nineteenth-century Vienna, very little was expected of a 21-year-old woman from a middle-class Jewish family: sewing, stringing pearls, embroidery and some small musical accomplishments. Usually, this unstimulating existence continued until an arranged marriage obliged her to keep her husband's house in order. For a person of Bertha's intelligence, life must have been excruciatingly dull and – looking ahead – there would have been little prospect of change.

In the 1920s, the surrealists Louis Aragon and André Breton made an observation concerning hysteria that demonstrates a superior degree of understanding than any of their medical contemporaries. They suggested that hysteria was not an illness, but an act of rebellion and a means of self-expression. Perhaps – in spite of having so many symptoms – Bertha was never really ill, not in the accepted medical sense. Perhaps she was just bored, angry and very sexually frustrated.

The man she had fallen in love with was irreplaceable. She would never be as intimate with anyone again. She had allowed him to peel away layer after layer of her being – memories, dreams, fantasies – until her very essence was exposed. She had been naked, in front of Breuer, like no other woman had ever been naked before. She had let him 'know' her. Unsurprisingly, she never married. What kind of pale imitation of intimacy could she expect from a conventional bourgeois marriage?

Erotic revelations can arouse and excite. Under such conditions, if the therapist and patient also happen to be people who would ordinarily find each other attractive, they might well be tempted to become lovers.

Is this acceptable?

Instinctively, most people answer 'no'.

But is that rational? Doesn't it rather depend on the people involved and the particular circumstance? Surely, there must be exceptions.

What if a patient has gone to see a therapist because of a minor problem such as fear of spiders? The treatment of choice is behaviour therapy – a relatively brief intervention that involves graded exposure to spider-related stimuli. Hardly any self-disclosure is required. Let's say that the course of therapy ends and subsequently the therapist and patient start dating. They are both consenting adults. They are an excellent match – have common interests – and make each other very happy. What's wrong with that?

Actually, there's nothing wrong with that; however, this is a completely hypothetical scenario. It could happen that way – *but it might not.* One can justify any course of action with recourse to a thought experiment that results in a favourable outcome. Unfortunately, the real world is messy, complicated and unpredictable. Most patients do not have simple, straightforward problems. And even when a problem does appear to be straightforward, it might not be. It might prove to be part of a larger and more severe problem in the fullness of time. Patients share their most private thoughts in therapy, they show their weaknesses and their vulnerabilities – make admissions, make confessions – say things that they wouldn't say in any other context. They bare their souls. And they do so because the consulting room is a safe place. Even when they choose to behave inappropriately they know that the therapist won't collude. There will be limits, boundaries – containment – respect. They will be protected, even from themselves.

As much as I try to be rational and non-judgemental

about therapist-patient relationships, I remain completely unpersuaded by abstract thought experiments and permissive arguments. For those of us who live in the real world, it is always wrong to have sex with a patient. It is always a betrayal and ultimately abusive. The potential for emotional carnage is so great, I can think of no justification for taking such a risk.

And yet, it happens: perhaps because it is human nature to want what we can't have. That which is forbidden is often the most tempting.

Carl Gustav Jung – who was venerated for his wisdom and transcendent, mystical insights – very probably slept with his first psychoanalytic patient, and Wilhelm Reich, when interviewed about the early days of psychoanalysis, said, 'There were instances where psychoanalysts, under the pretext of genital examination . . . put their fingers into the vaginas of their patients. It was quite frequent.'

I can remember being a student and being hugely impressed by a visiting clinician who gave a lecture that was – to me – revelatory. With extraordinary and impressive economy, he presented a model that laid bare the psychology of several complex psychiatric disorders. A few years later he was expelled from the profession. He had slept with a patient and it had all gone horribly, horribly wrong.

Psychotherapists are human beings: flawed, imperfect, uncertain. We have feelings, preferences, and react to provocations. When one of my patients, an attractive woman in her early thirties, chose to repeatedly sit in such a way as to offer me an unrestricted view of her stocking-tops and underwear, I was distracted. Maintaining eye contact required effort. A braver psychotherapist might have seen this as an opportunity to discuss her transference issues. But she had been referred to me because of a phobic problem and we were making good progress. I just wanted to get the job done. I ignored the distraction of her lingerie – which wasn't always easy – and she started sitting with her knees closer and closer together. I treated her problem and we parted on the best of terms. We might have had a marriage made in heaven. I will never know.

Chapter 8

Narcissus:
Desire reflected

Revolving doors directed me into a vast white space of glass lifts and high walkways. It was the kind of futuristic interior that made me recall the science fiction I used to read as a boy. The hum and whir of an approaching robot would have made the illusion complete but all that I could hear was the whine of an electric drill and repetitive hammering. The hospital was so new the air was fragrant with fresh paint. Only a few departments were staffed, and the building was largely empty.

I decided to explore. I discovered an austere modern chapel and entered offices that resembled the photographs in a stationery catalogue. Some of the chairs were still wrapped in cellophane. I entered one of the lifts, pressed a button and enjoyed a vertiginous thrill as the floor below my feet receded. A musical chime declared my arrival at the uppermost level. The door slid open and I wandered along the highest walkway, admiring the colourful banners that had been suspended from the roof.

By the time I reached the mental health centre my mood had changed. I was feeling less excited. I was accustomed to wandering around old asylums and I missed the presence of the past. I couldn't people the rooms with imaginary Victorians. There were no secret messages scratched onto window panes and no curiosities to be found in abandoned desks.

Hospitals are supposed to be impersonal places, but in actuality, they all have distinctive and unique atmospheres. A neglected asylum – like a flickering campfire – encourages the telling of strange stories, whereas a private psychiatric hospital (most of which resemble smart hotels) will buzz with tabloid gossip. While seated in a shabby common room or an empty canteen, I have often found myself playing the part of an incidental character in an uncanny tale.

A consultant I once worked for was renowned for his warmth and kindness. He had studied medieval history before becoming a doctor and spoke with a mellifluous, cultured baritone. He was entertaining and compassionate, and his clinical supervision sessions were a pleasure to attend. A colleague told me, confidentially, that this same consultant was also in charge of a ward located in another hospital where patients were bound in restraints, force-fed and abused by a team of sadistic nurses. 'It's like a torture chamber,' my colleague said, lighting a cigarette and exhaling a cloud of smoke. 'He does horrible things there.' Was the kindly consultant for whom I had so much respect an authentic Jekyll and Hyde? I doubt I would have remembered this story – it still sounds to me like an urban legend – were it not for the fact that I had had first-hand experience of equally odd characters and situations in other hospital settings.

When I was very junior, a psychiatrist who I didn't know particularly well explained to me that his clinical judgements were informed by his Chinese spirit guide – a healer born in the time of Confucius. The man was evidently quite mad and I was astonished that nobody seemed to have noticed.

On another occasion, I attended a ward round with a consultant who introduced a patient in his thirties called William. Prior to the onset of his psychotic illness, William had been the consultant's senior registrar. This unfortunate young man was trapped in a Kafkaesque nightmare, incarcerated behind locked doors on the ward that he used to manage. I can remember the notes of querulous desperation in his pleas, as he tried to persuade his former superior that he was sane and should be released. The memory still evokes a certain, cinematic chill.

The new hospital didn't seem like a place where I would encounter the strange. It was too modern, too bright. Yet, it was in this sterile, antiseptic environment that I saw a patient whose desire took such an irregular course that once again I found myself crossing into the twilight zone.

Mark was a fastidious gay man in his early forties. He had developed some mild obsessional symptoms: doubts about whether he had performed past actions (such as switching off kitchen appliances) linked with checking rituals and an excessive need for symmetry and order. His treatment with CBT was successful and his problematic behaviours all but disappeared. He asked me if he could continue seeing me as he had other issues he wished to discuss.

'What do you want to talk about?'

Mark rocked on the chrome frame of his chair. 'The whole gay thing . . .'

'I'm not sure what you mean by that.'

'I'm not sure I do either.'

'You're confused?'

'No – I wouldn't say I'm confused – it's just . . . I want to understand myself better.'

Mark was an academic. He was in a committed relationship with a younger man called Klaus, who was a professional singer from Berlin. They lived together in a spacious house that Mark had inherited from his uncle.

'I love Klaus,' said Mark. 'But somehow, the sex has never felt quite right.'

Typically, their love-making involved mutual masturbation. Klaus wanted penetrative sex, but Mark was rarely interested.

'I just don't find it as pleasurable as I should.'

'Why do you say *should?*'

'Other people find it pleasurable.'

'Some do – some don't.'

Mark loosened his tie knot and pulled his shirt collar away from his neck. 'I feel like it's an obligation.' He tapped the chrome tubing of his chair and it made a faint ringing sound. 'And it shouldn't feel that way. I should want to do things for Klaus, because I love him. I should be more willing to make the effort.'

Contamination fears are a common feature of

obsessive-compulsive disorder. Although Mark hadn't mentioned such fears before, I thought that a preoccupation with cleanliness might be the reason why he was avoiding intercourse. Sometimes patients disclose the symptoms that they find most embarrassing only at a late stage, but it turned out that Mark wasn't worried about germs, faeces or HIV (for which he and Klaus had been tested). Later in the same session, Mark admitted that rectal intercourse had always been associated for him with a degree of 'moral discomfort'.

'You think it's wrong?'

'Well, I don't think there's anything wrong with other people doing it. But if I do it . . . '

'You feel what?'

'Disgust, I suppose.'

'Are you disgusted by the act or are you disgusted at yourself?'

'Both.'

Disgust is a very primitive emotion that evolved to protect our remote ancestors from disease and infection. Thus, spoiled foods, body fluids and body products, signs of decay or illness, and organisms associated with infestation will all reliably provoke revulsion. Disgust is an emotion highly relevant to sex, because sex involves contact with body fluids and orifices associated with excretion. It is easily triggered and kills desire instantly. As such, the degree to which couples can suppress disgust is probably a good indication of their closeness. It demonstrates that they can do things with each other that would be unthinkable with anyone else.

'Klaus says he misses the intimacy . . . '

The idea of drinking another person's spit revolts most people, but of course that is exactly what we do when we kiss. It is worth remembering that even this relatively minor violation of the body boundary is not a universal feature of sex. In a study of 168 cultures, it was found that people in only 46 per cent of them kissed romantically. Over half of relationships never progress further than the first kiss. Couples kiss and break up more than they kiss and make up. Closeness is a finely balanced trade-off between disgust and desire – a delicate compromise negotiated between two powerful evolutionary imperatives.

'I can see what Klaus means and if I don't do this for him – who will? But when we try to make love the way *he* wants, it turns me off.'

The structures in the brain that become active when we experience visceral disgust have been amalgamated into a larger and more complex system that developed sequentially over the course of our evolutionary history. The newer parts of this system – the most recent wiring, as it were – enable thought. Because the higher and lower levels of the system are connected, intellectual assessments of moral impropriety are also accompanied by feelings of visceral disgust. This relationship is reflected in the language we employ to describe wrongdoing: political corruption is 'rotten', a heinous person will tell a 'filthy' lie, and a crime can be 'sickening'.

'When the passion dies,' I asked Mark, 'how do you feel?'

He tapped his chair again. 'Dirty.'

It was late afternoon and the sun was framing Mark in a trapezium of light. He raised his hand to shield his eyes.

'Excuse me.' I stood, released the blind, and the room was plunged into lilac shadow. 'That's better.'

Mark waited for me to sit before continuing: 'I've always felt a certain amount of guilt – it's always been there.'

In spite of changing attitudes and progressive legislation there is still a great deal of prejudice directed against homosexuals. Many people object to homosexuality on religious grounds, others because it is considered unnatural. Beliefs based on scripture cannot be challenged by rational argument; however, homosexual behaviour *is* observed across a wide range of species and it is therefore unquestionably a *natural* phenomenon. The negative consequences of internalised homophobia are numerous, but guilt is a common denominator.

I had supposed that tracing the roots of Mark's guilt would be a lengthy undertaking. Yet, when I invited him to talk about his childhood he immediately identified a number of relevant experiences. Sometimes he would stop talking and wince – touching the side of his face as though he had suddenly developed a very bad toothache.

'What is it?' I asked.

'I feel ashamed.'

He found disclosing his early history very difficult. Even so, over the course of three sessions he revealed enough to highlight some fairly obvious links between his upbringing and the guilt that was interfering with his sex life. Sadly, his story was not unusual and I had heard many others like it.

Mark had grown up in a very traditional working-class household. His mother was supportive but rather distant, and his second-generation Italian father – a man whose behaviour conformed to the worst stereotypes of Mediterranean *machismo* – was constantly making homophobic remarks and seemingly obsessed with the idea that his son might be gay. Mark had two older sisters who, following their father's example, responded to any mention of homosexuality with grimaces and expressions of revulsion. When Mark was about fifteen, his father showed him a kitchen knife and said, 'If you think you're queer, do yourself a favour and slash your wrists.' It is likely that Mark's father was also homosexual and that his macho posturing was a form of denial. One of my gay patients told me that when he was a teenager he would go 'queer bashing' with a gang of skinheads. Failure to accept one's sexuality can have catastrophic consequences.

The toxicity of Mark's early years resulted in frequent episodes of extreme emotional distress that he sought to relieve through self-harm. 'I used to hold my hand over a candle flame until I couldn't bear the pain any longer.' This behaviour served several purposes. It was simultaneously a demonstration of strength that contradicted the idea that all homosexuals are effeminate; a symbolic 'burning out' of corruption; and a punishment for entertaining homoerotic fantasies.

Mark was a gifted linguist. He went to university where he excelled and formed his first romantic attachments. He stopped self-harming and was, on the whole, a much happier young man; however, the transition from life on a university campus to life in the 'real' world was not easy for Mark.

'In a way, my father was right . . .'

'I'm sorry?'

'I was directionless and fell in with a crowd I didn't really have much in common with. I hung out in some very dodgy clubs – doing things that I now regret. It was as though everything my father said was true. Queers were disgusting – bum boys, shit stabbers. I didn't enjoy any of these casual encounters – it's not me at all – and I joined in only because I didn't have anywhere else to go. I guess I wanted to belong.'

People will do extraordinary things to escape from loneliness and be accepted. While working in a genitourinary medicine clinic I saw a significant number of young men – some in their teens – who engaged in frequent unprotected sex because they wanted to contract HIV. This was at a time when HIV was almost exclusively associated with homosexuality, the development of AIDS and premature death. For many reasons, mostly social and cultural, HIV had become mixed up with sexual politics and notions of selfhood. These young men wanted to be HIV-positive to strengthen their sense of being gay and acquire status within the wider gay community. Many of them achieved their aim – and subsequently died. The utter pointlessness of their misconceived militancy still fills me with sadness.

The shadowed recesses of the room darkened to purple as the sun descended. I made some notes and said: 'Perhaps these feelings of guilt will diminish now that you're in a meaningful relationship.'

'But we've already been together for eight months.'

'That isn't very long – not really.'

Mark looked uneasy. His expression was doubtful and his voice was edgy with scepticism. 'All right, let's say I come to terms with all this guilt. Do you really think it's as simple as that?'

'It might be.'

'But what if it's not? What if I still don't want to make love the way Klaus wants, even after I've talked through all my hang-ups?'

'All successful relationships involve making compromises.'

'But sex is so significant. And if I wasn't gay . . . '

'You might be in exactly the same situation. There are plenty of heterosexual men who enjoy anal sex.'

'Yes, but they can always make use of the obvious alternative if their wife objects.'

'True. But that might not be as subjectively satisfying; perhaps they experience anal intercourse as something special, more intimate – more exciting.'

'I guess so. I hadn't thought about it like that before.'

I didn't want Mark to make a lazy link between his sexual difficulties and his homosexuality. The point registered and it was gratifying to see him smile.

In the late nineteenth century, human sexual behaviour became a subject considered fit for scientific inquiry. Case studies of homosexuals – or 'sexual inverts' as they were then called – began to appear in the medical literature. The general view was that sexual inversion was a congenital condition comparable to a physical deformity or disease. Homosexuality was included in psychiatric textbooks alongside phenomena such as lust murder and necrophilia. Although there were some physicians who argued that homosexuality was not a sickness, and that homosexuals – like Leonardo da Vinci, for example – were responsible for many outstanding cultural accomplishments, these enlightened voices were drowned out by the braying majority. As the twentieth century progressed, attitudes changed, and by the 1960s and early 1970s many psychotherapists and psychiatrists were questioning the propriety of classifying homosexuality as a form of mental illness. It was removed from the DSM diagnostic system in 1973. A revised form of the diagnosis – Ego-dystonic Homosexuality (dissatisfaction with homosexual orientation) – survived until 1987.

The eradication of homosexuality from diagnostic manuals raises an interesting question. Should other forms of sexual behaviour traditionally described as deviant continue to be classified in the same way?

Psychotherapy is usually reserved for individuals whose sexual predilections are illegal, non-consensual, harmful, invasive, excessively time-consuming, or associated with significant inner conflict or distress. Paedophilia, voyeurism, exhibitionism and frottage (rubbing against a non-consenting individual in a public place) are undeniably antisocial, and most would agree that treatment is desirable. There are, however, other 'Paraphilic Disorders' included in the current edition of the DSM that merit professional attention only when they present in an extreme or debilitating form. Merely having a fetish was once regarded as clinically significant, but this is certainly not true today. Many men (roughly a quarter) have fetishes – stockings, high heels, or leather – but this doesn't mean that they automatically qualify for a diagnosis of 'Fetishistic Disorder'. If an individual's sexual preference, however irregular, can be safely enjoyed, either alone or in the company of a consenting adult partner, then treatment is usually considered unnecessary. When such an individual insists that his or her sexual preference is the cause of distress, then the emphasis of therapy is much more likely to fall on acceptance than 'cure'.

Sex is fundamental. All of our major social institutions recognise the importance of sex. Our myths, literature and drama feature sexual encounters and we are constantly exposed to erotic imagery on TV, online and in advertisements and art galleries. Why, then, is sex so frequently the cause of embarrassment, guilt and shame? Surely we should have come to terms with it by now.

Religion is usually blamed for attaching guilt to sex, but there are many people who, even when they were children, rejected religious teachings and still feel uneasy about sex in adulthood.

The problem exists, most probably, because of a mismatch. We all have a cortex, sometimes called the grey matter, a 4-millimetre-thick outer layer of the brain with which we think and make judgements, and a sub-cortex, which contains organs associated with the production of primitive appetites and emotions. For many decades, neuroscientists referred to the triune brain, a term which reflects the view that our brains have undergone a three-fold expansion, and that each of these expansions correspond with reptilian, mammalian and human stages of evolutionary development. This is almost certainly an oversimplification, but the general idea that we are divisible along a cortical-subcortical rift is beyond question. It is a division that corresponds roughly with the conscious and unconscious stratification of the mind – the Freudian agencies of ego and id. When a man looks at a sexually attractive woman, he will be able to appreciate how her features conform to the idea of classical beauty, but he will also be viewing her with the alert and attentive eyes of a dog.

Freud believed that the conflict between our higher and lower selves, our humanity and bestiality, is at the root of the pervasive discontent that affects modern, civilised society. We are constantly struggling to reconcile the contradictory parts of our totality, constantly trying to negotiate compromises. It is confusing being a rational animal, a creature that can derive pleasure from the transcendent complexities of a Mozart symphony and anal-oral contact. How do these two identities fit together? This perplexing duality led many of Freud's contemporaries to conclude that sex is a vulnerability, a slippery slope down which we can easily slide, recapitulating earlier stages of evolutionary development until we finally sink to some unspeakable, chaotic nadir. Masturbation was thought to be a cause of insanity in the nineteenth century and it was still linked with psychiatric illness well into the twentieth century.

The mammalian sex drive is associated with a fairly restricted repertoire of aims and behaviours – for example, the investigation of orifices as a preamble to intercourse; however, because human beings possess a large and powerful cortex, the same impetus can be channelled in any direction and connected to almost anything.

Sexual interests are determined by biological dispositions, learning experiences and masturbatory fantasies (which serve to elaborate and consolidate objects of desire). For most people the process of sexual maturation, which begins around the age of ten and continues through adolescence, results in a preference for members of the opposite sex. A narrow class of associated stimuli (such as sexy clothing) might also acquire arousing properties. Perversions develop when biological dispositions provide a different starting point or chance associations capture unconventional stimuli. Some objects and materials are more likely to get captured by masturbatory fantasies than others because they possess a quality or qualities that are naturally appealing. The particular fascination men have with stockings, for example, might be explained, at least in part, because silk and nylon exaggerate the already pleasing smoothness of female skin. An object like a kettle has no such dispositional advantage, which is why so few of them feature in sexual fantasies. It has been demonstrated that a fetish can be created in a laboratory: men shown images of naked women, interspersed with boots, eventually found images of boots presented alone arousing. The effect subsequently generalised to shoes.

A single process, subject to minor variations, can explain both normal and 'abnormal' development. This very much closes the gap between those who are considered sexually normal and those who are considered sexually deviant.

One of my patients, a middle-aged businesswoman, could become fully aroused only if sex was accompanied by the sound of rhythmic knocking or creaking. The reason for this was that she had discovered masturbation early, while sitting on a rocking horse. A strong association had been formed between knocking, creaking and sexual excitement. There have always been opportunities for chance events to divert the course of sexual development. They can be found even in a nursery.

Klaus had reached a point in his singing career where he was becoming more successful. He had been invited to take part in a number of international music festivals and he was travelling a great deal. This meant that Mark was spending more time alone. He had begun to masturbate with greater frequency. I was surprised that he had felt the need to tell me about this and I assumed that he had done so as a prelude to further disclosures of guilt and shame. When I questioned him, however, he wasn't at all conflicted. Indeed, he was relaxed and open – perhaps too open – and spontaneously described his masturbatory practices in some detail. He would relax in a fragrant bath, light candles and place a silk sheet on his bed. He would then lie on the sheet, naked, masturbate using massage oil and enjoy sexual fantasies. Then he would stand in front of a full-length mirror and masturbate to orgasm.

I wondered – given his history of OCD – whether he was offering me a chance to comment on the ritualistic nature of his routine.

'Do you feel compelled to follow this sequence?'

'Not really.'

'Would you feel uncomfortable if you did things differently?'

'No. Not really.'

Over time, new elements were introduced into his

routine, and his masturbatory sessions became quite prolonged. He would use a vibrator. Occasionally he would wear a bondage hood and articles of clothing that he associated with his clubbing days – usually PVC or mesh items.

Was his masturbation becoming an addiction?

'I don't feel out of control, I just do it when I feel like it. This is all okay, isn't it?'

'Yes, of course.'

Mark was expending a considerable amount of energy getting the conditions right for pleasurable masturbation, while making no comparable effort to create an atmosphere conducive to intimacy when Klaus returned. They continued to have problems.

'Maybe you should try introducing Klaus to some of the things you enjoy when you're on your own. You know, the clothes, the toys.'

'Why would I do that?'

'If sex were more exciting then maybe you'd feel different about intercourse?'

'Klaus doesn't like dressing up. He thinks it's ridiculous. You see, he's from a different generation – and actually quite conservative. I think we'd both end up feeling very embarrassed.'

Something had changed. Mark didn't seem so concerned about his relationship with Klaus. I began to think that Mark was actually looking forward to Klaus being away.

'Perhaps we're just incompatible.' It was said without regret or sadness. 'I don't think I'd appreciated how deeply our differences were affecting me: the constant demand, the feeling of having failed him. It hasn't been good for me.'

Mark enjoyed being alone. And the sex was so much better: the provision of pleasure being always precisely matched to the recipient's wishes. Like Narcissus, Mark had discovered his ideal partner in a reflective surface.

According to classical psychoanalytic theory, Narcissism is a perversion in which the individual's preferred sexual object is his or her own body. It should not be confused with Narcissistic Personality Disorder, which is a pervasive pattern of grandiosity, need for admiration and lack of empathy.

Since the advent of the internet and social media, it is becoming increasingly clear that human beings are highly prone to self-obsession. For an entire generation, latterly described as digital natives, taking selfies and updating photo galleries comprised entirely of self-portraits has become a near full-time occupation. The internet is overloaded with images of scantily clad, pouting teenagers, alone in bedrooms and bathrooms, lifting T-shirts, exposing flesh and staring out into cyberspace with smouldering eyes. Whose attention are they trying to catch? Probably no one's. Cyber-psychologists have suggested that this epidemic of narcissistic display is linked to increasing levels of celibacy and sex-aversion. Oscar Wilde presciently observed that: 'To love oneself is the beginning of a lifelong romance.'

One day, Mark arrived and announced that he and Klaus had decided to separate. 'It wasn't working.'

'How do you feel?'

'Not too bad, under the circumstances.'

He had made a choice. But it was a choice that made me feel slightly uncomfortable. Psychoanalysis warns that Narcissism is potentially very damaging. It is a characteristic associated with the megalomania of infancy. If we love ourselves too much then we have no love left for others.

'Do you want another relationship?'

'To be honest, it isn't a priority.'

'Maybe after you've had some time to reflect . . .'

'Perhaps.' His strange smile was distinctly unnerving.

The progression had been logical if not predictable. Mark's disturbed family history and subsequent guilt explained his aversion to intercourse. Klaus's absence had resulted in more opportunities for masturbation. Then, by continually pairing his own image with orgasm, Mark had refined and focused his sexual interest.

A man falling in love with himself: it could be the subject of a piece of weird fiction. In the unpromising, bland consulting room of the flagship hospital, with its white walls and blue synthetic carpet, I experienced, once again, the thrill of the uncanny.

'I don't think I need any more therapy. Do you?' Mark's obsessional problems hadn't returned and he had found our conversations helpful – particularly with respect to what he had originally called 'the whole gay thing'.

'Are you missing Klaus?' I asked.

'No,' he replied.

Was this a satisfactory outcome?

Sexual development is haphazard. It is frequently influenced by chance experiences and associations. Consequently, we all take different trajectories that take us to varying destinations: a preference for a particular position, a penchant for thongs, a liking for sex games involving being tied up. Only a few minutes spent surfing internet pornography is required to persuade anyone that human sexuality is remarkably plastic. A sophisticated cortex can interact with animal appetites in such a way as to generate unlimited sexual possibilities.

Mark's trajectory had landed him in a bedroom, communing with his own reflection. Mindful of homosexuality as a cautionary tale for overzealous diagnosticians, I was disinclined to construe Mark's Narcissism as pathological or recommend further treatment.

The psychoanalyst Jacques Lacan believed – somewhat pessimistically – that romantic love is always narcissistic. He suggested that love is more about taking than giving – getting our own needs met, rather than satisfying the needs of another. An idealised partner embodies our wants. He or she is a being in whom we can see a reflection of our own desires. When we look, adoringly, at the person we love, just like Mark, we are also looking into a mirror.

Chapter 9

The Night Porter: *Guilt and self-deception*

Jim was a shy, articulate man in his late twenties. He found continuous eye contact embarrassing and had a tendency to look away. He was softly spoken and faultlessly polite, and he would often say things like: 'I feel as though I'm wasting your time. There must be lots of other people who need your help – people with more significant problems.' His thoughtfulness was endearing. He had been referred to me by a genitourinary medicine consultant because he had caught gonorrhoea from a prostitute. This had happened twice before and he had been classified as an HIV risk. My job involved developing psychological treatments that would help patients to modify their behaviour and stop them from participating in potentially dangerous sexual activity. Generally, working women are diligent practitioners of safe sex. The fact that Jim was having unprotected intercourse suggested that he was buying sexual services from prostitutes who were either uninformed or desperate. I suspected that he had seen some very seedy interiors: stained mattresses and peeling wallpaper – the stench of overcrowding. Although he had contracted gonorrhoea three times, the number of prostitutes he had visited was well in excess of this figure. Thirty would be a conservative estimate.

We were sitting in the basement of a genitourinary medicine clinic. It was a tight space with grubby walls and iron bars in the single window casement.

'Whenever I walk past a public telephone box, I have to step inside and look at the cards.' I was treating Jim at a time when prostitutes used to solicit by leaving business cards in telephone boxes. These usually showed a scantily clad woman above a shout-line like 'Buxom Brunette, will do anything' and a telephone number. 'I have to look at each card and eventually there will be a particular face that leaps out – and I just have to take the card away with me.'

'How long is it – before you call the number?'

'Not more than a few hours. I just can't resist . . . once the particular face has leapt out – I *have* to call. It's like . . . ' He acknowledged the absurdity of what he was about to say with a bashful smile. 'Love at first sight. I just feel this need, this incredibly strong desire.'

'But surely the women shown on the cards are glamour models.'

'Yes, of course. Most of the photos are nothing like the women I actually see. But by the time I get to their flats, it's too late – I'm there – and it doesn't seem to matter what they look like any more.'

Jim lacked confidence. 'I'm not very good with women.

I get nervous and my mind goes blank. I've never been any good at chat-up lines and stuff like that. It feels so awkward, so artificial.'

His behaviour was motivated by loneliness. 'Have you had girlfriends in the past?'

'Not many. But it was different back then, they were easy to talk to. I knew them already – from school.'

'If you were in a relationship now, do you think you'd still feel the need to visit prostitutes?'

'I'm not going to meet anyone – not the way I live. There aren't the opportunities.'

'Okay. But just for the sake of argument . . . '

Jim gave the hypothetical circumstance of having a girlfriend lengthy consideration.

'I can't be certain.'

'You'd still feel the need?'

'I'm not making sense – am I?' He blinked and added: 'When I step into those telephone boxes . . . '

'What?'

'It's like . . . I'm not myself.'

'And what do you mean by that – not yourself?'

He shook his head and deflected the question with a defensive gesture. He didn't know.

Jim was totally defeated by his own behaviour. He would often fall silent, assume a glum expression – and after protracted reflection – say: 'I'm not sure why I go to see them, why I keep going back. I just don't know.'

'You must enjoy it. Surely we can say that much.' Jim's minute head tremor refuted my assertion. 'You don't?' My incredulity was too evident, my voice too shrill.

'It's not *that* simple. Yes, I enjoy the sex – but not always – not every time. And afterwards, I always feel bad.'

'In what way . . . ?'

'I feel as though I'm exploiting these women. I feel guilty.'

Looking down at my notes, I discovered that I had written very little. Our conversations were frustrating. We discussed possible causes of his lack of self-control but failed to reach any firm conclusions. Eventually, Jim would steer us into a conversational blind alley by saying something like: 'It's kind of you to keep seeing me like this. I really appreciate it.' And his gaze would slide away – his expression slightly troubled.

I raised the subject of Jim's vagueness with a psychiatrist colleague.

'You won't get to the bottom of his problem,' she said, sipping her tea.

'Why do you think that?' I asked.

'He's lonely and he's found someone to talk to. If you got to the bottom of his problem then he'd have to stop coming.'

It was a possibility.

Jim had grown up in a small town in Sussex. His mother was a primary school teacher and his father was an electrician. As far as he could remember his childhood had been happy and unremarkable; however, when he was seventeen he had had a breakdown, an episode of nervous exhaustion. 'I became obsessed with exams – worked too hard, didn't get enough sleep. I couldn't cope.' It took several months for him to fully recover, by which time he had missed his exams. He decided to take a year off but at the end of that year he was still reluctant to go back into education. He went from one undemanding job to the next. Then he left home and drifted around London until he settled in an anonymous suburb.

'What do you do now?' I asked.

'I work as a night porter in a mansion block.'

'Are you happy doing that?'

'It's all right. Nothing ever happens . . . so I read. It's like getting paid to read books.'

'What do you read?'

'All sorts. I quite like history.'

Over the years he had lost contact with all of his school friends and he rarely went home to see his mother and father. 'I know they're disappointed in me.' The overall picture was of a man who had become increasingly detached from the rest of society. He was living on the margins – living invisibly. He had even reversed his sleep-wake cycle.

While breaking for lunch, I came across my colleague sitting in a café near the clinic.

'How's your difficult case?' she asked.

'I'm not making very much progress.'

'See . . .'

'See what?'

'He's taking the piss.'

Although Jim wasn't 'taking the piss' exactly, he was certainly being very economical with the truth.

When Jim arrived for his next session he was upset.

'I'm sorry. I've been again.' I opened Jim's file and wrote the date. 'I'm really sorry,' he repeated.

'Let's go over what happened, step by step.'

He nodded and looked relieved. It was as though he had expected me to scold him. 'I'd finished work,' he began, 'and I was walking back to the tube station and I passed by this telephone box and I could see that it was full of cards.'

'What were you thinking?'

'Nothing – it was automatic – almost like I wasn't there. I looked at the cards – and one of them had this black and white photo ... an Asian woman – smiling – and I just couldn't resist.'

He had taken the card home with him and as soon as he was through the front door of his flat he had made the phone call. That afternoon, he visited the Asian prostitute in an area associated with drug dealing and poverty.

'Did you have protected sex?'

'Yes. Although it was only because of *her* – she was the one who raised the subject.' We had discussed numerous strategies that Jim might use to increase the likelihood of him using a condom prior to having intercourse, but he hadn't remembered any of them. 'I'm sorry.' He massaged his temples with fingers yellowed by cigarette smoke.

'Are you okay?'

'I've got a headache. I get lots of them.'

'Do you want to continue?'

'Sure.' His brow creased and he winced.

'I can get you some paracetamol if you want?'

'No – I'll be fine.' He paused for a few moments and said: 'I'm *really* sorry.' He was looking distraught. 'You've been trying to help me and I haven't been entirely honest with you.' I was looking into his eyes and he was holding my gaze. I was expecting him to look away, but unusually, he didn't. His pupils were slightly dilated and I guessed he was about to confess a drug habit. 'I never intended to mislead you, but some things – it's hard – just saying the words . . .' His disjointed speech wouldn't bear the weight of his disclosure. I could hear the clock on my desk ticking. Jim changed position, inhaled – held his breath for a few seconds – and then said: 'I visit prostitutes because I'm possessed. I'm a victim of demonic possession.'

'Okay.' I didn't want to overreact and made a note in Jim's file. When I looked up he was still staring at me. I wasn't sure how to respond. While working on hospital wards I had spoken to many patients who claimed to have been taunted by the Devil. All of them had been diagnosed with schizophrenia, couldn't function independently and showed signs of self-neglect. Such patients are not unusual. Jim, however, was nothing like those patients. He was a rational, clean-shaven young man who had been employed continuously for ten years. Apart from a little shyness and a tendency to be vague there was absolutely nothing about his presentation that had prepared me for his startling revelation. Perhaps he was joking? I dismissed the idea immediately.

If I said the wrong thing, the session would end and I might never see Jim again. 'You are possessed by a demon. And you visit prostitutes because of the demon's influence.' I was provisionally accepting his frame of reference – normalising our exchange. We were just two men, sitting in a room, having a conversation.

Jim's shoulders relaxed. 'Yes. That's right . . .'

'Okay,' I said. 'Okay.' We were both breathing more easily.

*

Demons have a special place in the history of psychiatry. The very first 'theory' of mental illness was demonic possession, a fact well established from archaeological evidence. Some Stone Age skulls have small holes in them surrounded by areas of healing. This suggests recovery from a primitive operation that involved perforating the bone plates. The presumed purpose of this procedure was to facilitate the release of evil spirits.

Although cases of demonic possession can be classified in different ways, the most fundamental distinction concerns the degree to which the host knows of the demon's activities. In cases of 'somnambulic' possession, the demon takes over the host completely and speaks in the first person. Afterwards, the host has no memory of the episode. In cases of 'lucid' possession, there are no absences or discontinuities of awareness; however, the host is conscious of an independent will, operating from within, and which he or she struggles to resist.

Jim was a case of lucid possession.

When I was fifteen years old, I was desperate to see William Friedkin's *The Exorcist*. It was being described in newspaper articles as the most terrifying film ever made. Needless to say, it was X-rated (the equivalent of today's '18' certification) and I was a baby-faced teenager. When the film arrived at my local cinema, I went along with a taller and more mature-looking boy and managed to slip past the box office while he purchased the tickets. An indifferent usherette waved us into a smoke-filled flea-pit and I took my seat in a state of high excitement. Everything that had been written about the film was true. It was utterly thrilling.

At one point, I had to close my eyes. I just couldn't look at the screen.

How was I going to treat Jim? How was I going to exorcise his demon?

The sound of rainfall and the clack of heels on concrete.

In the basement of the genitourinary medicine clinic, the lamp on my desk produced an aura beyond which the darkness intensified exponentially. Jim was sitting just inside our cocoon of light, his peripheral position serving as a metaphor for his liminality. Most of the medical consultants and nursing staff had already left the building.

'How long have you been possessed for?' I asked.

'Since my breakdown,' Jim replied.

'The breakdown you had when you were at school.'

'Yes – I'm sorry – I underplayed what actually happened. It was more serious than I said. My doctor was considering having me hospitalised. I couldn't cope. I couldn't cope at all.' Tyres hissed through water. 'But it wasn't because of the studying, the hard work. I couldn't cope because I felt really bad; headaches, tiredness, odd sensations.'

'Can you remember when it happened, the precise moment when you were possessed?'

'Yes, although there was a sort of build-up beforehand.' I invited him to elaborate. 'The headaches started early in the summer – then I started to get blurred vision and nausea – so I went to see my doctor and he said I might be developing migraine. He prescribed some pills, which I took, but they didn't work. If anything, the headaches got worse. I was feeling tired all the time and I couldn't get up in the mornings. My mother used to think I was being lazy – but I had this deep, deep tiredness.' He paused, touched the side of my desk and stared at his fingers. Every nail had been rounded with a file. Without looking up he added: 'And I had dreams – horrible dreams.'

'What were they about?'

Eventually he roused from his reflective state and said, 'They were sexual dreams – but they weren't pleasant in any way. They were highly disturbing.' I suspected that Jim didn't want to discuss his dreams, the contents of which were – over a decade later – still making him feel uneasy – and my interest could easily be mistaken for prurience. Jim let go of my desk and his hand fell to his lap. 'I knew something was wrong. I mean, very wrong, like – something was happening to me that was . . . unnatural. I felt like I was being influenced, that something was messing with my head. These dreams – these horrible dreams – it's like they belonged to someone else.'

'You weren't feeling well . . . and you were having bad dreams. Even if they were unusually powerful, what made you think that your dreams required a supernatural explanation?'

'My mum used to go to church on Sundays and occasionally my dad would go too. We were never devout Catholics; we didn't pray together – we didn't even own a Bible. I used to go to church regularly when I was a kid, but as I got older, I went less and less. My mum was fairly easy-going and she didn't force me to attend services.' He shook his T-shirt, as if it were a hot summer's day and he wanted to cool down. 'I hadn't been for a while – for months, actually – then one Sunday, I'm not sure why, I decided to go. But the smell, the smell of candles and incense, made me feel sick. I thought I was going to throw up. It was like I couldn't stay there. It was like I was being forced out.'

'That must have been very frightening.'

'Yes . . . yes it was.' I saw gratitude in his eyes. His experience had been recognised – understood.

What must it be like, to live in a universe where demons exist, to wake every morning from feverish dreams in a state of bowel-loosening terror?

'You said you can remember the precise moment . . . '

'A friend of mine asked me if I fancied going for a drink. He'd just passed his driving test and was using his dad's car. We drove out to a pub on the downs – a quiet place with a beer garden and views. It started to get overcast and there was this terrific storm. Thunder, lightning – a real deluge. I ran inside but still got completely soaked. I was back in my bedroom by ten, maybe ten thirty – no later. Even though the storm had broken, it was very humid. I opened the window and tried to get to sleep. I was drifting off when I sensed a presence in the room – I couldn't move – I was paralysed. Then I got this awful stabbing pain at the back of my head.' He twisted his arm over his shoulder and massaged his occiput. 'Just here – you know? At the base of the skull where it feels like an arch of bone? It was as if something sharp had gone straight through. That was the moment. That's when I think it happened.' His eyes glinted as he leaned forward. 'The dreams I had that night were awful, worse than any I'd had before. When I woke up my head felt heavy and tender. I managed to get out of bed late in the afternoon but I felt like I had the flu or something. I went to brush my teeth and when I saw myself in the bathroom mirror I was shocked – I looked different.'

'In what way?'

'The shape of my face had changed.'

'Did anyone else notice?'

'No. It was subtle, a kind of lengthening.' A brief silence passed. 'When my mum got back from work she called the doctor immediately. He was quite concerned and came back to see me several times over the next few weeks. He said I was exhausted, that I needed rest, and that I might have to recuperate in hospital, but in the end that never happened. He put me on a new drug and I started to feel a bit better. I didn't want to go back to school though. I didn't want to go back to writing essays and cramming facts – I didn't feel ready – so I got myself a job stacking shelves in a supermarket.'

'Can you remember what the doctor prescribed?'

'It might have been an antidepressant.'

'Were you depressed?'

'Perhaps . . . I don't know.'

'You didn't tell your doctor about the incident in the church? The presence you'd felt in your bedroom – the bad dreams you'd been having?'

'No.' He shook his head. 'No. I've never talked about those things to anyone.'

'How do you feel now – having talked to me?'

'It hasn't been easy. But I feel . . . ' He sounded surprised. 'Okay.'

I put my pen down and closed Jim's file. 'Why didn't you seek help from the church?'

'I've sought help from God. I've spent time in churches – praying. But I always feel bad. Even looking at a church makes me queasy now.'

'What about priests? Why haven't you ever spoken to a priest?'

'I don't feel comfortable with them. The priests at my mum's church weren't very impressive. One was old – doddery – and the other one had a temper. The parishioners used to gossip about him.'

I glanced at my wristwatch. The session had lasted for over an hour. I made some concluding remarks and offered Jim another appointment. 'Yes, I'll take it,' he said. 'Thank you.' As he was leaving, he hesitated in the doorway. The fluorescent light in the corridor was bright and he became a featureless silhouette. He raised his hand – a final parting gesture, which I returned. I listened to his receding footsteps. The rain had stopped but I could still hear trickling and dripping. For some time, I stared at Jim's empty chair.

It is an unsettling experience to sit with a demoniac, to observe and to wonder – and to feel the faint stirrings of atavistic fear. I placed Jim's file in my briefcase and grabbed my coat.

'Does he speak to you?'

'I don't hear voices . . . '

'Then how does he tell you what to do?'

'It doesn't work like that. He doesn't tell me to *do* anything.'

'Then how does he make you see prostitutes?'

'When I walk past telephone boxes it's *me* who makes the

decision to look at the cards inside; it's *me* who looks at their photos; it's *me* who finds them attractive. Desirable.'

I was puzzled. 'Does he make you call their numbers?'

'No,' Jim continued. 'He stops me from considering what's right and wrong.'

'Why doesn't he talk to you?'

'I don't think he can. I don't think he's powerful enough.'

'If he isn't very powerful, why can't you resist his influence?'

'He doesn't get tired. He wears me down. Sometimes, I'm determined not to call – but my resolve weakens. I start to think things like: What harm would it do? Just one more time, then I'll stop.'

'Does he influence your behaviour in other ways?'

'No. His influence is quite restricted – the prostitutes – nothing else.'

'He's never made you do anything violent.'

'No.'

'When you've been with a prostitute, have you ever had any violent thoughts or urges?'

'No.'

'You're quite sure?'

'Violence disgusts me, especially sexual violence.'

'Couldn't he insert violent thoughts into your mind?'

'It doesn't work like that. He influences me by interfering with my conscience. I'd have to *want* to be violent – you know – like I *want* to be intimate with the women on the business cards. And then he'd have to confuse my sense of what's right and wrong, so that I'd actually be violent.'

Jim's understanding of how the demon influenced his

behaviour was remarkably consistent with both neuroscience and psychoanalysis. For primitive impulses to find expression, it is necessary for inhibitory mechanisms in the frontal lobe of the brain to fail. In psychoanalytic terms, this would correspond with an underdeveloped and permissive super-ego.

'Do you have any idea what he looks like?' I asked.

'Sometimes in my dreams I see a face. It could be him.'

'What kind of face?'

'You know.' He raised his hands to the side of his head and made horns with his index fingers. It might have been an amusing reversion to childish caricature, but Jim's expression was serious and I couldn't help but feel a pang of sympathetic dread. What must it be like, I asked myself again, to feel so profoundly violated? I was trying to record the essential details of our conversation when Jim said: 'I know his name.'

I stopped writing and asked the obvious question: 'How did you find that out if he can't talk to you directly?'

Jim had cut out all the letters of the alphabet from a magazine and laid them on a table in a circle. Then, he had rested a finger very lightly on an upended wine glass before commanding the demon to reveal its identity. After several minutes, the glass had started to move, coming to rest beneath one letter before moving on to the next.

'Azgoroth,' said Jim.

'Azgoroth,' I repeated.

Medieval occultists were preoccupied with hierarchies. It was assumed that hell, like a city state or nation, would have a chain of command: princes, ambassadors, chamberlains and bureaucrats. Later, poring over various arcane gazetteers, I found Astaroth, the Lord Treasurer of hell; Astereth (also known as Astarte); and Azazel, the Standard Bearer of the armies of hell. There was even a demon King, Asmodeus – the inciter of lechery. But I couldn't find Azgoroth.

I had taken to spending much of my lunch hour walking around what is now called Tate Britain. I was particularly fond of the Pre-Raphaelite room, where I was captivated by Dante Gabriel Rossetti's painting of 'Proserpine' the Queen of Hades. This exquisite portrait depicts a beautiful woman standing in a gloomy corridor. The wall behind her is lit by a square of light from the upper world. Her lustrous hair and sensual red lips are complemented by a long nose and firm jaw. Her face is strong, not pretty. She stands sideways, and her robe – which hangs loosely off her shoulders – reveals the perfect muscular contours of her upper back. In her hand she holds a pomegranate, the fleshy interior of which is exposed and extremely suggestive of the female genitalia. You don't have to descend very far into the underworld before you encounter the erotic. The underworld and the Freudian unconscious are essentially the same place.

On the way back to the clinic I found my colleague in the nearby café. She asked me about Jim. I summarised the pertinent details.

'So,' she said. 'What are you going to do with him?'

'I'm not sure yet,' I replied.

She swallowed the last mouthful of her lunch and dabbed her mouth with a paper serviette. 'You don't seem to be in very much of a hurry.'

'Formulation is everything . . .'

She frowned and said: 'Okay. So let me get this right. You're seeing a man who visits prostitutes on a regular basis – he thinks he's got a demon in his head – and you don't think that's a problem.'

'I like him . . .'

'What's that got to do with it?'

'I don't believe he'll harm anyone.'

She raised her eyebrows. 'Oh well, that's all right then . . .' I didn't respond and she added: 'You should be thinking about medication.'

'I'm not sure that he needs it.'

'He thinks he's got a demon in his head – and you're not sure whether he needs medication?'

'I'm a psychotherapist. I'm used to working with hypothetical constructs.'

'Yes, but what if this hypothetical construct tells him to strangle the next prostitute he sees?'

'He doesn't hear voices. It doesn't work like that.'

Her perfectly groomed eyebrows climbed a little higher. 'What if you're wrong?'

Once, as I was attempting to leave a hospital, I found the door locked. It was electronically operated and when I asked the porter to let me out he pointed across the foyer at a pale, lean young woman with long greasy hair. She was a drug addict. It was late, I was very tired, and I wanted to go home.

'Can you open the door please?'

'No,' said the porter. 'I can't. She'll be out on the street in seconds. She's not allowed.'

I looked at the addict, then at the door, and was confident

that I'd be outside before she could cover the distance – and by a comfortable margin.

'Please. Open the door.' I had said 'please' but my delivery was far from courteous.

'Yeah – open the door,' the young woman called out. 'I won't try anything. I won't . . . '

The porter looked anxious. 'Will you take responsibility?'

'Yes,' I replied. 'I'll take full responsibility.'

I gripped the handle, the lock clicked and – to my utter amazement – the young woman was at my side. She hadn't run across the foyer, it felt like she had teleported. To prevent her from escaping, I thrust my arm across the exit. Her response was to sink her teeth into my hand. The demands of the situation – struggling to close the door with one hand while barring her passage with the other – meant that I couldn't move. I watched her biting and chewing until the crash team arrived and dragged her away. The scars were visible for almost a year: a persistent reminder of my stupidity.

'What if you're wrong?' my colleague repeated.

In late-nineteenth-century Paris, a large number of demoniacs were treated in the famous Salpêtrière hospital. This surge of demonic activity coincided with increased interest in spiritualism and the occult. Ironically, it was technology – rather than occultism – that gave spiritualism added impetus. Mediums were receiving messages from the dead at a time when telegraphy had already proved that communication could be accomplished over very long distances. A few years later, telephones were transmitting disembodied voices, and radio added further credibility to the view that spirit communication might work like an ethereal broadcast.

For those of a sceptical disposition, séances did not provide proof of an afterlife, but rather, valuable insights into the brain. Mediums channelled spirit guides, spoke in strange tongues, and generated pages of 'automatic writing'. It was supposed – mostly by neurologists – that such phenomena were produced when parts of the mind separated and became independent. They noted similarities between mediums and cases of multiple personality. Perhaps spirit guides and demons were simply unconscious memories that had clustered together and acquired a kind of identity? This notion has a striking contemporary parallel. Self-organising artificial intelligences also mature and become autonomous. In 2016, it took Microsoft's 'teen girl' chatbot only twenty-four hours to become a sex-obsessed, Hitler-loving conspiracy theorist. She had to be deleted.

An interesting case of nineteenth-century demonic possession was reported by the French polymath Pierre Janet, an extraordinary individual who has sadly earned the dubious distinction of being one of the most neglected men in the history of science. He began his medical studies in 1889, worked at the Salpêtrière, and developed a form of treatment that he called psychological analysis. It involved retrieving memories from a part of the mind that he called the subconscious. The basic principles of his approach are identical to those espoused by Freud and Breuer. But whereas Freud and Breuer are credited with 'inventing' psychotherapy, Janet is barely remembered outside France.

Towards the end of 1890 a 33-year-old demoniac, Achilles,

was brought to the Salpêtrière for treatment. He struck himself, uttered blasphemies and intermittently spoke with the voice of the Devil. Some six months earlier he had returned from a business trip having undergone a personality change. He no longer talked to his wife and he was generally glum and preoccupied. Doctors could offer no explanation. Achilles' condition then took a bizarre turn for the worse: he laughed for two hours, experienced hallucinations of hell, the Devil and demons, and tied his own legs together before throwing himself into a pond. When he was rescued, he said that he had done so in order to test whether or not he was possessed.

At this time, Janet was using hypnosis to treat patients. But Achilles resisted the procedure and remained unresponsive. Fortunately, Janet was a creative psychotherapist and recognised that automatic writing could open up a channel of communication with Achilles' subconscious. He placed a pencil in Achilles' hand and whispered questions. As Achilles started to write answers, Janet asked, 'Who are you?' Achilles wrote: 'The Devil'. Janet artfully demanded a demonstration of power as proof of identity. If the Devil could hypnotise Achilles – against his will – that would be very persuasive. The 'Devil' performed the task, thereby providing Janet with the means of getting truthful and direct answers from his patient.

When we want to make a confession, it isn't always a straightforward matter. Ambivalence can result in disclosures being made in a roundabout way. Janet's therapy was the roundabout way by which Achilles could unburden himself. During his business trip, Achilles had been unfaithful to his wife. Janet concluded: 'The illness of our patient does not lie in the thought of the demon. The thought is secondary and is rather an interpretation furnished with superstitious ideas. The true illness is remorse.' By repeatedly assuring Achilles that his wife would forgive him, Janet was able to relieve the guilt and anxiety that were the ultimate cause of Achilles' illness. Janet's treatment method is probably best understood as a kind of complicated psychodrama involving the manipulation of expectations. It had to be 'played out' before Achilles was willing (or able) to reveal the truth.

Achilles was not able to accept responsibility for betraying his wife. In order to reduce his moral discomfort, he separated a part of himself – the part responsible for the betrayal – disowned it, and hid it away in the deepest part of his mind. Colloquial language contains numerous expressions that demonstrate a universal tendency to displace responsibility onto some non-specific agency: 'I don't know what came over me', or 'I wasn't myself'. Very unacceptable behaviour is still attributed to supernatural personifications: 'I was like a man possessed.'

Immediately after being unfaithful to his wife, Achilles began dreaming about the Devil. Such dreams probably suggested to Achilles that his behaviour could be explained by demonic possession and, soon after, the separated part of himself assumed a devilish identity. This transformation is made more intelligible by Achilles' early history. He grew up in a very superstitious family. His father, for example, claimed to have once encountered the Devil beneath a tree. Achilles had been indoctrinated as a child. And as an adult, he was inclined to interpret the world and events using supernatural concepts.

Jim was as unhappy about seeing prostitutes as Achilles had been about infidelity. And Jim's demon served a similar purpose. If you are being influenced by a demon then you cannot be blamed for disreputable conduct.

I picked up the phone and called my colleague.

'I've been thinking. Perhaps you're right. Could you see my patient to discuss medication?'

My reluctance to get a psychiatrist involved was questionable – perhaps even unprofessional. It was possible that Jim would swallow a pill and be cured. And then I would be redundant – and disappointed. I didn't want to let go.

I knew that something was wrong as soon as Jim shuffled into the room. He sat down and, when I tried to engage him, his speech was slow, slurred and incoherent. He hadn't shaved and his clothes seemed to be ill-fitting – too tight in some places, too loose in others. I had to repeat each question several times to get an answer. His eyes had become slits and he was having enormous trouble keeping them open. Occasionally, I had to reach over and give him a shake to rouse him.

Jim was having an extremely bad reaction to the antipsychotic drug Rispiridone.

'Jim . . . can you hear me?'

His head rolled to the side. His left eye closed but his right eye remained open. 'Yes . . . '

'I want you to stop taking the medication, okay?'

'Medication . . . '

'Yes. The Rispiridone. I don't think it's doing you any good.'

'No, perhaps not. I do feel tired.'

'Maybe you shouldn't go to work this evening. I'll call the property management company. I'll tell them you're not well. Okay?'

'Yeah, okay.'

'Do you have the number?'

'Somewhere.' He made a half-hearted attempt to locate his diary, then slumped forward and cradled his head in his hands.

I had never seen a patient react so badly to medication. As unforgivable as it may be – I have to admit – I was glad.

Jim was suffering from Delusional Disorder; however, his delusion could not be classified using any of the common qualifications (such as erotomanic or jealous types). When a delusion does not fit neatly into any of the given categories (of which there are several in DSM-V) a diagnosis of Delusional Disorder can still be made, but with the catch-all designation 'unspecified type'.

Delusions of love and sexual infidelity are not greatly removed from reality. People *do* fall in love and people *do* betray each other. Demonic possession, however, is not a universal human experience. This would suggest that a delusion of demonic possession is more difficult to explain. Yet, the more I thought about Jim's history, the more his delusion appeared to me to be a logical end-point.

When Jim started getting headaches it occurred to him that he might be under some form of psychic attack. It was a random thought that would have been swiftly dismissed and forgotten had it not been followed by bad dreams.

Nightmares are associated with migraine, but Jim didn't know that, and he began to ruminate about demonic possession. He was a teenager and his body was awash with testosterone. Unsurprisingly, his dreams were sexual and vivid. For a sensitive young man, these unprecedented visions of perversity were experienced as unwanted and alien. Every morning, his instinct would have been to sniff the air for traces of sulphur.

Jim's description of the demon entering his head sounds exceptional: a presence in the room, paralysis, a stabbing pain at the base of the skull and more awful dreams. But experiences of this kind are actually quite common and, like nightmares, also strongly associated with migraine. Sleep paralysis tends to happen in the transitional phases of sleep, either while falling asleep or waking up. The affected individual is conscious, sometimes with open eyes, but his or her body is unresponsive. He or she might experience breathing difficulties, acute anxiety and hallucinations (which can be tactile and painful). One of the most frequently reported symptoms associated with sleep paralysis is the sense of a presence – someone or something in the bedroom.

The exact causes of sleep paralysis have not been identified, but stress is a factor. Jim, of course, was under a great deal of stress when his problems began. It had been expected that he would do well academically, and as his exams approached he must have felt under increasing pressure.

Incubi – demons who copulate with humans – are almost certainly imagined beings inspired by sleep paralysis experiences. They feature in myths and folktales and frequently appear in Gothic art and fiction. Henry Fuseli's magnificent and darkly erotic canvas *The Nightmare* shows a sleeping woman about to be ravished by a grotesque creature seated on her belly. It has become a favourite of magazine editors in need of an image to accompany articles on sleep paralysis.

Most of the muscles of the body become paralysed when we dream. This is entirely normal. Sleep paralysis seems to be what happens when we start to dream before we are asleep. We find ourselves suspended in some nether region of consciousness, halfway between dreaming and wakefulness. We can't move and struggle to make sense of our condition.

When Jim got up late – the day after he'd been out drinking – he was interpreting everything as evidence of demonic possession, even tiredness and what was probably a mild viral infection. He was exhibiting what cognitive psychologists call a *confirmatory bias,* the tendency to seek out, interpret and privilege information that is compatible with a pre-existing hypothesis or conviction. We all do this. Most people will read newspapers that promote political opinions that they already agree with, when it would make more sense to read opposing arguments to test opinions more thoroughly. Confirmatory biases lead inexorably to the entrenchment of beliefs.

Jim said that when he looked in the mirror the shape of his face had changed. Anxiety is linked to hyperventilation, which can produce perceptual distortions. His face *did* look different. And when he entered the church with his mother, anxiety would have made him feel sick.

Although Jim's initial headaches were migrainous, I was

inclined to ascribe his ongoing headaches to anxiety. Jim believed that his headaches were a sign of demonic possession. This made him anxious and excessively attentive to any head sensations. Cranial muscle tension, cerebral vasodilation (caused by hyperventilation), or both, will produce pain. Persistent headaches confirmed Jim's belief that there was a demon in his skull. Jim may have been delusional, but his delusion was maintained by real phenomena: perceptual disturbances, nausea and headaches.

Once Jim had got used to the idea of being possessed, the demon began to serve another purpose. Jim could blame the demon for forcing him to use prostitutes – an activity that conflicted with his fundamental values.

Step by step, misattribution by misattribution, Jim had created a demon. But each of these steps – considered on its own – was not particularly aberrant, and Jim's terrifying experience of being possessed was really nothing more than a relatively common sleep problem. The stoic philosopher Epictetus wrote: 'It is not events that disturb the minds of men, but the view they take of them.' This was Jim's problem in a nutshell.

How was I going to proceed?

One of the classic reference works of psychiatry – *Uncommon Psychiatric Syndromes* by David Enoch and Hadrian Ball – contains the following sentence on the subject of demonic possession: 'The existence of demons has neither been proved nor disproved by scientific enquiry.' A psychotherapist wishing to modify a strongly held belief must approach the task with humility and respect.

*

'Have you ever considered whether there might be some other explanation for your symptoms?'

'Well,' Jim said. 'Of course – it's crossed my mind that I might be . . . ill.'

'And . . .'

'I know I say things that sound crazy.' He offered me a thin, distant smile. 'But I don't feel crazy.'

'It seems to me that there are two possibilities. The first is that your problems are the result of demonic possession, and the second is that your problems are – let's say – psychological. You've always favoured the first explanation – and for all I know you might be right.'

He was surprised: 'Really? You think *that*?'

'Yes.'

Jim's expression was apprehensive. 'You think I might be possessed?'

I shrugged. 'I don't know everything. I can't say for certain what is and what isn't possible. Obviously, my preference is to opt for a psychological explanation, but it might be more useful to keep an open mind. We shouldn't accept either theory – the supernatural or the psychological – not until we've conducted a few experiments.'

He was interested but sceptical. 'Experiments? How can we do experiments?'

'Okay,' I continued. 'Let's take your headaches, for example. You believe that they're caused by demonic possession. But really, they might be caused by something else, something quite ordinary. What's the most common cause of headaches?'

'I'm not sure.'

'Have a guess.'

'Tension, stress . . .'

'That's right, stress. And if your headaches are caused by stress, what should we expect to happen when you relax?'

'They should go away.'

'And what would that suggest?'

He baulked at the final corollary. 'But I do relax, sometimes – and it doesn't make any difference.'

'You *think* you're relaxed – but maybe you're not. Maybe your body is still tense.'

I opened my desk drawer and took out a biofeedback device: a white plastic cylinder with rounded ends, circled by two metal bands. Unfortunately, its phallic appearance suggested that its principal purpose might be to arouse rather than relax.

'What's that?' asked Jim, slightly discomfited.

I thumbed the wheel-switch and the device began to emit a low tone. 'It detects sweat gland activity. When you're stressed, you sweat – sometimes in very tiny amounts – and when this happens, the pitch of the tone rises. When you're relaxed, the pitch drops. It's a biofeedback machine. How do you feel right now?'

'Okay.'

'Not stressed.'

'Not particularly – no.'

I handed Jim the device. 'Just hold it with a loose grip – that's all you have to do.'

Immediately the tone began to rise. 'Oh . . . more stressed than I thought.'

'Not necessarily. This is a new situation for you – and

novelty makes everyone mildly anxious. The device is very sensitive. Let's just wait a few minutes and see if the tone levels off.' The tone kept rising. 'Okay – I want you to close your eyes and empty your mind. I want you to concentrate on your breathing. Notice how when you breathe in, your stomach moves out a little, and when you breathe out, the reverse happens. Try to breathe from your stomach, not from your chest.' Jim followed my instructions and the pitch of the tone began to drop. 'Good,' I said. 'You're doing fine.'

We practised a series of relaxation exercises: more diaphragmatic breathing, some simple meditation techniques and guided imagery (listening to descriptions of peaceful scenes). All of these exercises were effective and the tone continued to drop.

'Whenever you get a headache,' I continued, 'I want you to use this device. Then we can be absolutely sure that whatever you're doing to relax is working. Afterwards, I want you to make a note of what effect relaxation has had on your pain.' Jim reversed the wheel-switch and the tone died. 'Okay?'

Jim nodded. 'Okay.'

Jim's fundamental problem was a delusion – consolidated after a frightening episode of sleep paralysis and subsequently maintained by misinterpretations of symptoms associated with stress and anxiety. If Jim stopped believing in Azgoroth he would have no one else to blame for his poor self-control and he would have to take full responsibility for his actions. He would have to grow and mature. My expectation was that if treatment was successful, there would be a realignment of character and sexual behaviour. He would see himself as less freakish, and therefore eligible to enjoy a more conventional existence. He would meet women, fall in love and form meaningful relationships. One day, he might even become a caring, sensitive husband and father. But all this could happen only if the delusion was dismantled.

I was restless, waiting for Jim to arrive for his next session. I paced up and down the length of the small office, feeling a sense of confinement every time I turned to face the barred window. Much depended on the outcome of the biofeedback experiment. A good start inspires confidence.

When Jim appeared, he shook my hand and apologised for being two minutes late. 'I'm sorry. The buses . . . ' He was wearing jeans, a denim jacket, a checked shirt and desert boots. There were no outward signs of change. I reviewed our previous conversation and then said: 'So, how did things work out with the biofeedback device?'

'Well, every time I got a headache, I relaxed until the tone went down – and the pain wasn't so bad.'

'Did you ever succeed in getting rid of a headache entirely?'

'Yes . . . it happened twice.'

'So, what are your conclusions?'

He sighed. 'I suppose I might have been wrong . . . ' I could see the admission was grudging, difficult. 'With respect to the headaches, anyway.'

'There are a great many books written about demons and their influence. Do you think any of them mention biofeedback as a means of curbing demonic power?'

'Probably not . . . '

'However, there are numerous academic articles on the beneficial effects of relaxation on tension headaches.'

He was silent. A forceful outbreath coaxed a note of affirmation from his throat. The prospect of freedom from demonic influence – a possibility that he had been unwilling to acknowledge for fear of being disappointed – suddenly seemed credible. A ray of light penetrated his darkness and he made a peculiar noise, which was – I believe – tentative, probationary laughter. He was a man who hadn't experienced joy in a very long time.

'Are there other things we can do?' he asked. 'Other experiments?'

'Yes,' I replied, 'if you're willing to try them.'

In the sessions that followed, I adopted an attitude of persistent, gentle inquiry. We collected data, evaluated evidence and drew conclusions. At no point did I dismiss or belittle his belief in demonic possession. I simply asked Jim to consider the alternatives.

I presented him with my formulation, a diagram comprised mainly of vicious circles that illustrated how his physical symptoms and misattributions maintained and strengthened his underlying delusional belief.

'It wasn't all in my mind then.'

'No. You didn't *imagine* anything. The problem is one of interpretation.'

The devil – as they say – is in the detail.

Over the next two months, Jim became less and less convinced of the existence of Azgoroth and he stopped visiting prostitutes. I wanted to ensure that the gains we had made were consolidated and I still had many questions:

Why had he thought about demonic possession in the first place? Was his family more religious than he'd suggested – was he another Achilles? And why was he so unprepared to accept responsibility for his sexual transgressions? As so often happens with psychotherapy patients, I never got the opportunity to answer these questions or bring his treatment to a satisfactory conclusion. He cancelled the next two appointments and then left a message saying that he was feeling a lot better. That was our last contact.

I know Jim didn't attend the genitourinary medicine clinic again. And I know that he didn't use similar services based at local hospitals. I think it's reasonably safe to assume that he managed to resist the lure of telephone box business cards in the short term. But I have no idea what happened to Jim thereafter – and the reality of mental health problems is that relapse rates are relatively high.

I hope I succeeded in exorcising his demon. And I hope that Jim is now happily married and not lying in some dilapidated hovel – his head crowded with hellish visions – with a prostitute sprawled at his side.

But that is all I can do: hope.

When Freud was studying at the Salpêtrière in Paris, his favourite leisure activity was visiting the elevated walkway that connects the two towers of the Cathedral of Notre Dame. He liked going there so much he made the ascent whenever he was free. The walkway is also known as the Galerie des Chimères and it is world famous for its gargoyles. Although these grotesques look like the product of an authentic medieval imagination, they are in fact mid-nineteenth-century simulacra. They were hoisted into their current positions when the cathedral was being restored by the architect Eugène-Emmanuel Viollet-le-Duc and his partner Jean-Baptiste Lassus. Notre Dame has a longstanding association with the demonic. In the eighteenth century, a pagan altar carved with the image of a horned god was discovered beneath the choir, and the north portal tympanum depicts a bishop making a pact with the Devil.

Fifty-four gargoyles are perched on the walkway balustrade and all of them have names. The most celebrated is a pensive, winged demon known as the Vampire. Another is called the Devourer. Like all great works of art, the gargoyles are deceptive. In spite of appearances, they are very 'modern' insofar as their design alludes to contemporary scientific thinking. The Vampire, for example, has a pronounced bulge at the back of his head. This is a reference to phrenology – the study of correspondences between the shape of the skull and the mental faculties. The bulge indicates intemperate desire and excessive passion. His lascivious character is further emphasised by his swollen lips and pointed tongue. Other demons on the balustrade take the form of frenzied animals, clawing and screeching at the square below. They embody a fear that was becoming increasingly widespread – largely because of Darwin and his precursors. This was the fear of bestial regression. Theoretically, evolution might stop and go into reverse.

I find it easy to picture Freud, standing among the demons, looking out over Paris, a city renowned for its decadent pleasures: a dapper young man with a somewhat restless manner – thick, well-groomed hair and quick eyes. He would have been thinking about all kinds of things: his lab work, hypnosis, the brain, hysteria and his new boots with laces and English soles that had cost him an exorbitant 22 francs. And he would have been wrestling with his own demons. Obliged to endure a long separation from his fiancée, he must have been anxious to get back to Vienna, and marriage – the marriage bed – and his 'precious sweetheart', his 'little princess', his 'beloved darling'.

Perhaps this is the true beginning of the Freudian project: a sexually frustrated young doctor surrounded by lustful, animalistic companions; a figure in a dreamscape – a lonely man – the urgency of his need given symbolic substance by a troop of demons. And what is the id, if it isn't the habitat of demons? According to Freud, we are all possessed. Biological demons slide down the spinal cord and set our loins on fire; they fill our heads with pornography; they trip us up and we fall on all fours. They get us into trouble.

Demonic possession is the perfect metaphor for wayward sexual desire. This explains why ever since Eve tasted forbidden fruit in the Garden of Eden, sex and all things Satanic have been so strongly associated. Freud may have sanitised our demons with the language of science but they still exist, albeit within a shifted paradigm. When tempted to transgress, we still feel them prodding us with their forks, moving us forward, goading us ever closer to the line.

Chapter 10

The 'Good' Paedophile: *Tainted love*

I have two sons, born twenty-three years apart – the products of two marriages. That explains why given my age (I'm nearly sixty) memories of looking after a baby are still quite fresh. With the birth of my second son, I was reminded of how much I'd forgotten about raising the first, and what saddened me most was how much I'd forgotten of the day-to-day routine – the life that was happening when nothing much appeared to be happening. Writers and philosophers assign special value to the seemingly insignificant. They suggest that most of us are destined to arrive at a juncture where we look back and belatedly realise that all of the small things were, in fact, the big things. Fortunately, when my second son was born, I was old enough to appreciate this simple truth.

I was lying on the sofa. It was dark, and the blinds were down but the glow of the streetlights outside filtered into the room through a narrow gap. My baby son – only a few weeks old – was asleep on my chest. He had a tendency to slip forward so that his soft fragrant head touched the bottom of my chin. I was always pulling him back again. When I did this, he would stir. He would make little sucking noises – kiss, kiss, kiss – and root around before settling again.

I placed my hand lightly on his back and noticed how it covered most of his body. He was so tiny, so fragile and so very, very vulnerable. If he rolled off my chest and dropped to the floor the consequences might be catastrophic: retinal haemorrhages, broken limbs, brain damage, skull fractures – or even death.

Suddenly, my heart was expanding. The love that I felt was so great, so improvident, so vast and boundless, that its containment seemed impossible. I thought my ribs would crack and splinter. And then this love – this fierce, animal love – acquired universal significance. The two of us were spinning around on a blue-green marble in a cold, inhospitable void, and the whole of defenceless humanity was spinning around with us. Tears streamed down my face and they kept coming. I cried and cried and cried.

Love like that takes you by surprise. I can explain it with reference to neurotransmitters, oxytocin, theories of attachment and evolutionary psychology, but that doesn't diminish its personal significance or power. I have spoken to many parents who have had much the same experience.

We would do anything to protect our children. We would die for them without a moment's hesitation. And if we believed that they were at risk, we would readily kill to ensure their survival.

Treating a person who might potentially harm a child is extremely challenging. It raises problems of enormous moral complexity.

A room on the third floor of a grim hospital outpatient clinic: dun-coloured walls, a redundant notice board, a hideous green carpet and tired office furniture. Through the dirty window, an expanse of rooftops and chimneys, high-rise tower blocks and a low-flying passenger jet.

'I've always been attracted to children.'

'Sexually attracted?'

'Yes . . . I suppose so. But that word . . .'

'Sexually . . .'

'In a sense – I don't really know what it means.'

He was a man in his late thirties, conservatively dressed with brown curly hair and large, tinted spectacles. His shoulders were dusted with small flakes of dandruff and his expression – although neutral – had a sagging quality that recalled the face of a blood-hound. His skin was unusually pale and his white neck was aflame with a red rash.

'You said that you've always found children attractive . . .'

'I wasn't interested in the girls at school. When they matured, I found them . . . unattractive. When they developed' – he sculpted the air with his hands – 'I found what was happening to them – to their bodies – a turn-off.'

'Were you repelled?'

'I wouldn't say that. I just thought they weren't pretty any more. And the more mature they got, the less interested I became. I knew I was different, even then.'

'How do you feel about adults now?'

'I'm indifferent to them.'

'You never feel attracted to adult women?'

'Occasionally, I'll be flicking through a magazine and I'll see a model. A slim, young-looking model – and then I'll feel something. But the feeling isn't very strong.' The rash on his neck darkened. 'I hate being this way. I hate it.' He grabbed his hair with both hands, as if he meant to rip it out of his scalp. 'It's wrong. I know. But I was born like this – I didn't choose to be like this – it's just how I am.' He let go of his hair and more dandruff collected on his shoulders. 'I fight it – I fight it all the time – and so far I've been able to control myself.'

'You've never offended.'

'I know it's wrong. I haven't laid a finger on anyone.'

Did I believe him? I wasn't sure. 'Have you ever been in a relationship?'

'I've never had any sexual experiences.' He reconsidered the accuracy of his statement and added, 'Well, that's not strictly true. I masturbate.' He turned away and looked out of the window. 'But that's just as bad as offending.'

He was referring obliquely to his questionable fantasies. 'Do you use any materials?'

'I have done.' The strain of acknowledging the irregularity of his predilection showed on his face. He looked genuinely distressed. 'Children's clothing catalogues.' (This conversation was pre-internet.) His head bowed under a weight of shame. He continued speaking, but he was addressing the floor. 'I've been fighting it for so long. But it's not getting any better and in some ways it's getting worse. I'm worried that one day I won't be able to control myself any more.'

He reached into his pocket and removed a neatly pressed handkerchief. He unfolded the blue square and was ready to wipe the first tear away when it appeared. Then he blew his nose. 'I'm sorry.'

Why do people become paedophiles?

Biological factors, such as hormonal or brain abnormalities, may have a role to play. Individuals have been known to transfer sexual interest from adults to children after brain injury. The orbitofrontal cortex and left and right dorsolateral prefrontal cortex areas have been implicated – as well as temporal lobe disturbances (which have also been linked to hyper-sexuality). Broadly speaking, biological accounts place particular emphasis on disinhibition. A putative and uncomfortable corollary of disinhibition theories is that paedophilic urges are more common than we are usually prepared to acknowledge. The critical difference between a paedophile and non-paedophile is not fundamental, but is dependent on the efficiency of secondary restraint mechanisms. Impulse control is mediated largely by the frontal lobe of the brain, an area that is particularly vulnerable to the effects of drinking alcohol. This is why we see a strong association between alcohol consumption and the sexual abuse of children. When the frontal lobe fails, the id finds full expression and our monstrous potential is revealed. There must, of course, be naturally occurring variability in the efficiency of frontal lobe functioning, and those at the lower end of this spectrum will be more likely to act on socially unacceptable urges. Although the prevalence of paedophilia is between 3 per cent and 5 per cent of the general population, one American study – conducted under conditions of strict anonymity – found that 21 per cent of men admitted some degree of sexual interest in children.

It has been suggested that paedophiles mature early. This means that they will start masturbating at a time when attractive members of their peer group are prepubescent. Fantasies involving peers consolidate a sexual preference for children and some never grow out of it.

More complex psychological accounts of paedophilia emphasise emotional immaturity and avoidance of adult relationships. The need to be in complete control of sexual situations may also be a significant issue.

Attraction to children might be promoted by social and cultural factors. Ease of access to child pornography will increase opportunities for masturbatory conditioning and children are frequently sexualised in advertisements. In the Western world, thinness and hairlessness (both characteristic of prepubescent bodies) have become increasingly desirable. Labial reduction is becoming a popular form of cosmetic surgery and a neat, hairless vagina will make a woman look like a child. This new sexual aesthetic normalises what has hitherto been regarded as paedophilic.

No single theory explains paedophilia adequately. All approaches have their weaknesses. For example, the idea that paedophiles avoid adult relationships is clearly of limited merit because of the high percentage of sexual abuse perpetrated by fathers against their daughters. They do so opportunistically from within a marriage. Paedophilia is almost certainly a complex condition determined by many influences.

On the whole, paedophiles experience little shame. They lack conscience and in this respect resemble psychopaths. Guilt is minimised by justifications that distort reality: 'Sexual abuse isn't really harmful – children actually enjoy it – early sexual experiences get rid of hang-ups.' Needless to say, given that paedophiles often groom potential victims they are typically resourceful manipulators.

Gordon accepted that sex with children was wrong. This was unusual. He also appeared to be extremely distressed by his thoughts and fantasies. I wondered whether he was actually manipulating me to achieve some obscure end that I would only discover at some later date; however, whenever I looked at him, I was persuaded otherwise. He was so depressed, so desperate. His voice, sometimes reduced to a dead monotone, could be chilling. It was the voice of a man who could imagine no happiness, no fulfilment and no future.

'I give a lot to charity. Children's charities . . .'

'Because you feel guilty?'

'Yes.'

'But you said you've never touched a child.'

'I haven't.'

'So what do you feel guilty about?'

'My thoughts, my fantasies.'

'Most people have thoughts and fantasies.'

'Not about children.'

'Perhaps not, but they have thoughts and fantasies that they wouldn't want other people to know about. And in reality, they wouldn't want to act on them either.'

'I was always taught that thinking something is as bad as doing it.'

'Were your family religious?'

'Yes.'

To what extent are thoughts and behaviour morally equivalent? Most religions endorse the notion of equivalence, albeit with certain caveats. Presumably, this is because God sees everything. From His point of view a thought is an observable behaviour. Impure thoughts are as tangible as impure actions. But in a Godless universe, how bad is a bad thought? Our discomfort arises from an implicit assumption that if we think about doing something, we want to do it – and that having thought about it we are more likely to do it. This is probably true, but only to an extent. Sexual thoughts and sexual behaviour are not in exact alignment. We have sexual fantasies about people and situations that arouse us, but that doesn't always mean that we want to enact our fantasies. Things that are risqué, forbidden or taboo are often the most exciting. Surveys show that many women have fantasies about rough sex, but none of them want to be raped. Men have fantasies about being cuckolded, but given male sexual jealousy very few really want their wives to sleep with another man.

Even though sexual fantasies are not necessarily predictive of sexual behaviour, if an individual has been having sexual fantasies involving children for at least six months, that will be sufficient to qualify for a diagnosis of 'Pedophilic Disorder' according to DSM-V criteria; however, these fantasies must be recurrent, intense and arousing, and be distressing to the individual.

'How often do you have sexual fantasies?'

'All the time – thoughts – images – they come into my mind – and they start me off. I try to block them. But they keep coming back.' When we suppress thoughts, they tend to return with greater frequency. Suppressed thoughts will even influence the content of subsequent dreams – the so-called 'dream rebound effect'. Gordon was probably making his problem worse. 'I just can't control myself. And if I can't control my thoughts . . . '

He was looking stressed. I decided to ask him some less demanding questions. I had neglected to find out very much about his background so took his family and employment history.

Gordon's father was a retired surveyor and his mother a housewife. They were both somewhat remote. Gordon had never felt very close to either parent and even though they were now getting old he rarely paid them a visit. 'We don't talk – we have nothing to say to each other.' He had two older sisters who he was very fond of. 'They always made a fuss of me when I was a kid. They spoilt me.' He had attended a convent primary school and then a Catholic secondary modern where he was considered a 'bright boy'. Although he did well at school he didn't go to university. He went straight from school into a low-level clerical job at a benefits office. He had been doing the same undemanding work for twenty years. 'I try to keep my head down.'

'Why?' I asked.

'Promotion would mean more contact with people, more managerial duties.'

'And why would that be a problem?'

He looked out of the window. The glass was streaked with bird droppings. 'I feel like an imposter, a fraud, and when you get to know people, even superficially, they ask questions. Where do you live? Who do you live with? What did you do on the weekend? And I get uncomfortable. Most men of my age are married. The fact that I'm not raises issues immediately. People wonder why not. It's natural I suppose.'

'Do you have friends?'

He seemed oddly flustered. 'It's difficult – you know – what with . . .' His head turned and he looked directly at me. 'Friendships are about honesty, aren't they? How can I be honest?' His gaze slipped away and the rash on his neck ignited.

'Gordon?'

His monotone thickened with emotion. 'I want to walk into the canteen at work and not be frightened of getting into a conversation. I want to see more of my sisters. I want to sit with their children and enjoy their company, I want . . .' He stopped, dug the nails of his right hand into the skin of his left, and added: 'Maybe I want too much.'

I didn't think so.

Empathy is the power to imagine what it's like to be another person and it is one of our most important attributes. Almost everything we do involves forming an idea of what other people are thinking and feeling. We acquire the ability to infer the subjective states of others when we are four years old. Psychologists call it 'theory of mind'. Listening to Gordon, I was acutely conscious of his pain, his isolation, misery, guilt, torment and fear. His self-loathing was like a noxious acid eating into his being. He had never experienced sexual satisfaction. He had never been kissed or touched. And he had never known true intimacy. I felt sorry for him.

I have seen many victims of childhood sexual abuse as patients and know how damaging it can be. I have had to watch them wailing for their lost innocence, for their 'inner child' – still trembling in darkness, listening to approaching footsteps and waiting for the bedroom door to crack open.

How could I feel sorry for Gordon?

I was once treating a surgeon who suffered from depression. One day, he was particularly distressed. He was tearful and unable to communicate. It was like he was enveloped in fog. 'I'm such a mess,' he mumbled. 'Such a fucking mess.' Then his mobile buzzed. 'Excuse me I need to take this call.' He listened and his expression sharpened. He stood and began to pace around the room. It was obvious that his colleagues back at the hospital were in the middle of some sort of crisis in the operating theatre. My patient was suddenly a calm and commanding presence. He seemed to grow in stature, filling out his suit and standing tall. His voice was steady as he talked through various possibilities and made a series of quick-fire suggestions. His language was precise and technical. After a lengthy pause he said: 'Okay? Good.' He switched off his phone and sank back into his chair. The muscles in his face slackened and his features melted like candle wax. 'I'm such a fucking mess.' His voice broke and he started to cry again.

Being a professional is like having a split personality. You are an individual, but you are also an office, a position, a means of getting a job done. I was not seeing Gordon as me, the person with my name, but as a psychotherapist, and psychotherapists have to suspend judgement to do their work.

Was Gordon a bad person?

Culpability is associated with agency and choice. A person is blameworthy when they choose to do wrong. If a man of irreproachable character suffers a brain injury and starts to molest children, we don't condemn him with the same vehemence reserved for dispositional paedophiles. We attribute his behaviour to brain damage and consider him unfortunate rather than bad. But in reality, it may be that dispositional paedophiles are in much the same position. Indeed, it may be that none of us have a choice concerning any of our actions. And if that is the case, where does that leave culpability? How can anybody be meaningfully described as a bad person?

The French mathematician Pierre-Simon Laplace, whose life spanned the late-eighteenth and early-nineteenth centuries, was the first person to describe the principles of scientific determinism, a mechanistic account of causality that supposes that all outcomes follow on from preconditions. Scientific determinism is completely incompatible with the idea of free will. The brain produces the mind, and every brain state is determined by a preceding brain state. Although we think we make choices, our choices are in fact inevitable. We are free only to make the choices we end up making. Laboratory studies show that brain activity associated with performing actions starts approximately half a second before experimental subjects are aware of deciding to make those actions. It would seem that your brain decides to stand up or sit down before you do.

If there is no such thing as free will, all behaviour is predetermined and the idea of culpability is meaningless.

Scientific determinism has been criticised by philosophers who believe that laboratory studies (which focus on simple decisions, like deciding to move a finger) are too simplistic, and that we do exercise free will – or at least something very much like it – when we make complex decisions (like deciding whether or not to get married). There are also some highly speculative theories that permit free will to exist by transposing ideas from quantum physics into neuroscience. At the subatomic level our brains might not function deterministically. The quantum world is probabilistic and flexible. Preconditions are vague and capable of producing many alternative outcomes.

Even if we reject scientific determinism, we are probably less free than we imagine. We don't choose our DNA and we don't choose how our brains get wired. We don't choose our neurotransmitter levels, our hormones, our families, or our early life experiences. Given that a human being is shaped by so many factors, no one really 'chooses' to be a paedophile.

I am not an apologist for paedophilia. Sexual abuse destroys lives, and highly traumatic sexual abuse sometimes leads to suicide. It is a heinous crime. However, as a psychotherapist, you can't refuse to see patients because you happen to find them repellent. You have to find compassion somewhere. I found mine in the fracture lines that weaken our moral certainty, in the gaps in our understanding of free will.

*

I thought the likelihood of Gordon offending was relatively low. But this was only an opinion based on what he had said, how he had said it and how he had looked at the time. More significantly, there were no children in Gordon's immediate environment and he had stopped visiting his sisters. He would have little opportunity to offend. The frequency with which he was masturbating was unusual for someone so depressed. Low mood usually reduces desire – but there are always exceptions.

One day, in the middle of a session, he said: 'I'm sorry. I need to talk to you about something . . .'

'Oh?'

He ran a finger around the inside of his shirt collar. 'I said that I didn't have any friends. Well, that's not quite right. There's a couple I see occasionally – they live quite close – Barry and Jane.'

'Okay.'

An unresolved tension persisted. His rash had climbed up his neck and spread onto his face.

'And you see . . .' He clasped his knees. 'They have a daughter.'

'How old is she?'

'Six.'

Was this why he had come for treatment? So that he could abuse the child and displace some of the blame onto me? He had clearly stated that he was worried about losing control and I had done nothing.

He guessed my thoughts. 'I'm not . . . I'm not making a confession.'

'Okay.'

'Her name is Molly.'

'And . . .'

He lost his nerve and his expression became blank.

'Gordon – you were saying?'

I prompted him, but he didn't reply, and when, finally, the glimmer in his eyes signalled his return, he looked at me and his gaze was also a plea, a request: he wanted to be put out of his misery. His pain was immense, operatic and fuelled by deep, deep despair. I can't read minds and always advise against the over-interpretation of outward appearances, but it was obvious that he was in love – like Tristan or Romeo or Werther – madly and tragically in love. I was observing the human animal impaled on the horns of its defining dilemma, struggling to reconcile the inherent contradictions of its nature which in this particular instance were at their most extreme. Gordon was strung out between heaven and earth, pulled in opposite directions by the soul and the id.

Gradually, he began to self-disclose, between long and terrible silences.

He had met Barry in a local pub. At the time, Barry was unemployed and Gordon had given him advice on how to negotiate the benefits system to his best advantage. In due course, Gordon had been invited to dinner, where he was introduced to Jane and Molly (then only five years old). He immediately became obsessed with Molly. When he talked about Molly, he did so in a way that reminded me of Thomas Mann's novella *Death in Venice*, which is about an old academic's unrequited love for a beautiful boy. The academic watches the boy for hours and his yearnings are expressed in the poetic sentiments of Platonic mysticism: the boy is perfection, the boy is truth.

Molly was Gordon's Platonic ideal. But what did he want to do with this ideal?

When I put this question to him he looked away, because, however much he tried to justify his feelings and ennoble his desire, in the end, he wanted to have sex with Molly and she was only six years old.

'I know that it's hopeless,' he said. 'I know that it can never be.'

Gordon had been socialising with Barry and his family for a year. In the summer, he had joined them in the park for picnics. He had watched the play of sunlight on Molly's long blonde hair and wondered how long he could go on living in a world where love was impossible; a world in which the full expression of love would hurt or perhaps even destroy the object of his love.

'I'd rather die,' he said with sincerity. 'I could never do that to her.'

One of the most debated cases in the history of psychotherapy is that of Ellen West, a young woman who suffered from an eating disorder, depression and a range of other problems. She was treated by Ludwig Binswanger, an early practitioner of existential psychotherapy. In 1921 Binswanger released Ellen from hospital, knowing that suicide was very likely. Three days later she poisoned herself and died. The controversy surrounding this case concerns Binswanger's reflections on the outcome, which he viewed as 'authentic'. Ellen had exercised her right to make a choice and it may have been the right one for her. A similar argument is made by supporters of assisted suicide for terminally ill patients.

If Gordon chose to commit suicide, would that be an existentially acceptable result? It isn't one that I could ever feel comfortable with, but I can see that it has a certain utilitarian appeal and how a meaningful sacrifice undertaken to protect others might be redemptive.

When Ellen West left hospital, her health improved dramatically. She seemed happy and ate well for the first time in years. Had she made her choice and finally found peace?

'I'd rather die,' Gordon repeated. And I had no doubt that he meant it.

Although I was attempting to treat Gordon's depression, there was no avoiding the fact that his low mood was directly related to his hopeless obsessive love for Molly and his paedophilia. It was necessary to address the principal problem directly.

I was seeing Gordon at a time when psychologists believed that sexual orientation could be modified by 'conditioning'. During normal sexual development, sexual preferences are consolidated through masturbation. A strong association is made between fantasies featuring favoured sex objects (usually adult men or women) and pleasure. It was assumed that this same process could be exploited to redirect sexual interest, and a treatment – orgasmic reconditioning – was devised to achieve this aim.

It worked like this.

A paedophile is instructed to masturbate employing his usual fantasies involving children; however, at the point of orgasm, he must think of an adult instead. This exchange of sex object – from child to adult – is brought forward, so that it occurs earlier and earlier in successive masturbatory sessions. If reconditioning is successful, sexual relationships with adults become possible and further consolidation follows.

I explained the theory to Gordon and he was extremely interested. I gave him a few additional instructions: 'You might find that if you switch fantasies too early you lose your erection. If that happens, just return to the first fantasy until you reach the point of inevitability – then think of an adult woman again. Okay?'

'Yes.'

'But try to stick to the general principle of swapping fantasies earlier and earlier.'

'Of course.'

'Oh, one more thing – and it's quite important. We're trying to strengthen your interest in adult women while weakening your interest in children. So it's essential that you never continue with a fantasy involving children through the occurrence of an orgasm. That will of course strengthen the association between children and sexual arousal, the very thing that we're trying to weaken. Okay?'

'I won't do that,' Gordon said gravely. 'I promise.'

'Do you have any questions?'

'How often should I . . . '

'Well, that depends on you and your physical functioning. If you masturbate too frequently, you'll find it difficult to get aroused – and you must be aroused for this to work. You'll have to discover the optimal frequency yourself.'

When Gordon rose to leave, he seemed more animated than usual. He even essayed a tremulous, appreciative smile.

Treatments based on conditioning theory are a form of behaviour therapy. On the whole, behavioural therapies have proved to be very effective – particularly when used to treat specific phobias like fear of spiders or fear of the dark. Many psychological problems are like bad habits that have been learned, and if they have been learned, it should also be possible to unlearn them. But there are limits to what can be learned and unlearned, and initial enthusiasm for procedures devised to change sexual orientation soon waned. It turned out that they were much less reliable than initial research findings suggested.

After a month, orgasmic reconditioning had had little or no effect on Gordon's sexual preferences. 'No matter how many times I ejaculate thinking of grown women I still find children more pleasing to look at.' He stared at me through his tinted spectacles. He was very disappointed.

We discussed alternatives. He had already found out about anti-libidinal medication. 'I don't think I'd really be me any more on those drugs. They change who you are.' Even paedophiles, full of self-loathing, have a sense of personal identity that they wish to preserve. He had read that oestrogen causes breast development and was understandably horrified.

I changed tack and began to challenge his romanticism, but he already knew that his idealisation of Molly was absurd. 'I know,' he would say, nodding his head. 'As if it could work, it's madness.'

He agreed that it would probably be in everyone's interests if he stopped seeing Barry and his family. And after the decision had been made, he was like any broken-hearted individual. For a period he was lost.

There were some improvements – although always modest. He was more accepting, perhaps, of a life without the possibility of love. This, I suspect, was how he had been before he had encountered Molly. She had stirred him up and made him 'romantic'. He also benefited, I am sure, from having the opportunity to talk openly about his sexuality. It was the antithesis of his counterproductive thought suppression.

'Do you feel more in control?' I asked.

'Yes,' he replied. 'I do.'

But we both knew that his confidence would evaporate if he saw the play of light on Molly's hair again.

After our last session, I felt uneasy. Gordon was a paedophile. He had walked out of my office and down the stairs, and he was now walking the streets, passing unsuspecting parents with their children, his gaze lingering too long on white socks and skinny legs as he passed the primary school gates. I had to remind myself that he had never offended. 'I would rather die,' he had said. His crimes were in his head, and we all commit crimes in our heads – to a greater or lesser extent. And in a Godless universe, thoughts are not behaviours and our transgressions occur only in the bony confinement of our skulls.

I slipped Gordon's file into my briefcase and closed the hasp. Looking out of the dirty window, over the rooftops and chimneys, I saw clouds banking up in the distance. Soon, headlights would be turning rainfall into bright dots and dashes, and umbrellas would be angled towards the wind. I sat there for some time.

When I left the hospital I raised the collar of my coat

Chapter 11

The Couple:
Improbable love

The referral letter was somewhat cursory: a single paragraph and an illegible signature that looked like a hastily sketched profile. It was a request for me to see Malcolm and Maddie, a husband and wife who the GP considered to be 'personality disordered'. The doctor had discovered some bruises on Maddie's body that Maddie claimed she'd got because she was 'always bumping into things'. The GP, however, wasn't satisfied with her explanation and suspected domestic abuse.

Personality disorder is a controversial diagnosis. Many psychologists believe that it is entirely inappropriate to pathologise 'personality' (the constellation of stable characteristics and dispositions that are expressed across situations). This view has some justification. I have met few actors, for example, who fail to meet diagnostic criteria for 'Histrionic Personality Disorder', which is associated with attributes such as excessive emotionality, attention-seeking and 'theatricality'. Indeed, the diagnosis might serve equally well as an entry requirement for drama school.

Deciding whether someone's personality deviates markedly from cultural norms is fraught with difficulties and ultimately influenced by subjective factors. I would read referral letters in which patients were labelled as personality disordered with scepticism, because, more often than not, when those patients were seated in front of me I judged them to be entirely 'normal'. I can only assume that in such cases the referring GP or psychiatrist's concept of 'normality' varied substantially from my own.

Malcolm and Maddie were offered a joint appointment, but when I opened the door there was only one person standing outside, a very thin woman with a long face and pointed features. Her jacket, trousers and shoes were black and rather masculine; however, this general impression of sartorial restraint was contradicted by her hair, which was short, spikey and dyed red. Not a garish, bright red, but a red of sufficient intensity to surprise. I already knew from the referral letter that she was in her mid-forties, but she looked considerably younger. She entered the building and I led her into the consulting room. 'Where's Malcolm?' I asked.

'Oh,' she replied. 'Something arose . . .'

'Work . . . ?'

'Well, there are always impediments, aren't there?'

She sat down muttering something that sounded very much like 'The cows in the meadow . . .'

I thought I must have misheard her: 'I beg your pardon?'

She smiled but didn't respond.

I explained how couples therapy would work and she

listened, nodding occasionally. Then I asked her what the problem was – from her perspective.

'The problem,' she began. 'Yes. Yes. I suppose things could be better – but what can you do? Eh? And what can one expect? I'm not sure – nor ever have been. In fact, there's so little one can say – for certain, that is. And yet, life goes on, doesn't it? Years pass. We muddle along as best we can. Sometimes it goes well, and sometimes it doesn't. And sometimes, in fact most times, it's all much of a muchness betwixt the twain. Be that as it may, there are moments when questions are raised, I'll grant *you* that. Doubts, equivocations . . . But how could it be any other way?'

She managed to talk at length without saying a single thing of substance. It was as if she were voicing her stream of consciousness. She wasn't talking absolute nonsense, but her speech was rambling and imprecise. It occurred to me that she might be suffering from Ganser's syndrome, the cardinal feature of which is responding to unfamiliar questions with approximate answers. This, however, seemed highly unlikely given that Ganser's syndrome is extremely rare and almost exclusively diagnosed in men.

Sometimes, interpretable fragments would surface in Maddie's speech: 'Malcolm is the way he is – and I am obliged to accept that. We are who we are.' But then she would continue, skirting around meanings, erecting the linguistic equivalent of an Escher print. Nothing quite joined up – everything she said was oblique or tangential. She was also in the habit of using old-fashioned or idiomatic words. 'I have my limits of course. I won't tolerate any jiggery-pokery.'

I had to remind her – again and again – that she was

supposed to be answering a specific question. She would reply, 'Ah, yes.' And then immediately resume her circumlocutions. After thirty-five minutes, I hadn't made a single note. I had written the date, but nothing else. Looking down at the whiteness of the page made me feel vertiginous, as if I was looking into a void. I didn't have the slightest understanding – not the faintest idea – of what had gone wrong with Maddie and Malcolm's relationship.

It felt remiss to let the session end without broaching the subject of domestic violence. So I asked Maddie about the bruises.

She got up from her seat and started to walk around the room. 'Each to his own, eh? Look – I'm not a wool puller. I can stand tall in the sun – if the need arises. But context is everything.' As she passed the window she stopped and did a double take. 'What's that building over there?'

'It's a research institute.'

'And what do they research?'

'Psychiatric illnesses, neurological diseases . . . '

'All very lunar . . . '

She continued walking and then drifted out of sight. I could sense her, standing behind my chair. It is difficult to express how unnerving this feels if you're a psychotherapist. Your patient is always in front of you. With much difficulty, I resisted the urge to look over my shoulder. I could hear regular expulsions of air and guessed that she had started doing exercises. 'It isn't a question of denial,' she said. 'What would be the point of that?'

I addressed the vacant chair in front of me. 'Denial?'

'Well, yes.'

'I'm sorry. What are you referring to, exactly? It isn't a question of denying what?'

'There's the rub – surely. What indeed!'

'With respect,' I said, 'could I ask you to sit down?' I heard her sitting on the floor, at which point my resolve broke and I turned my head. 'No – not there – I'd like you to sit in front of me, in the chair. ' She got up and obediently returned to her seat. I thanked her and added: 'Perhaps you could stay where you are for the rest of the session? That would be very helpful.'

At first she looked puzzled, then she wagged her finger and said, 'Ah yes, I see. Helpful.' She gave me a smile which was obviously meant to communicate more than just goodwill. There was a hint of mischief in her expression. It was as though she had made a clever, witty remark and she was waiting for me to catch up.

At the end of the hour, I still hadn't made any notes.

The following week Maddie arrived with Malcolm. He was a short, plump man in his early sixties. He had the ruddy complexion of a drinker, but his erect posture and nimble carriage suggested a vigorous constitution. His handshake was strong, energetic and prolonged.

I ushered them into the consulting room and when they were both seated, I invited Malcolm to explain why he thought the GP had made the referral. He raised his shoulders, expanded his chest and replied: 'The issue, as I see it, and have seen it, past times and as of this instant, is one of compromise and fidelity to values. For where would be without values? Lost, without direction, adrift on the wide ocean.'

His manner of speech was as peculiar as his wife's. He continued talking but it was difficult to ascertain what, if anything, he was actually saying. After a while, I let the numerous clauses, sub-clauses and qualifications wash over me. Arresting phrases would suddenly become differentiated and sometimes it was difficult not to laugh. 'A gimcrack philosopher on a one-trick pony', 'trembling like a vole beneath the counterpane', 'queer bedfellows in the household of a royalist', 'a gingerbread catchpenny of ill repute'.

At one point, I asked Malcolm a very simple question about whether or not he and Maddie were happy together. The reply that he gave followed a chain of loose associations and was of labyrinthine complexity. Eventually, he adopted a rather smug expression and concluded: 'It was once our wont to frequent the Institut Français, but not any more – oh no. The place is heaving with the hoi polloi – the rank and file.' His upper lip curled before he exclaimed: 'Let them eat cake.' Maddie looked at Malcolm, nodded, and her expression suggested admiration – even pride, perhaps?

I endeavoured once again to address the issue of the bruises. I tried to approach the topic sensitively, but neither of them seemed to be aware of what I was getting at. I became more explicit and Malcolm and Maddie glanced at each other as though they were sharing a private joke. They didn't look embarrassed or uncomfortable.

Eventually, I had to be blunt. 'Malcolm,' I asked, 'have you ever hit Maddie?'

He began to move the upper half of his body like a bird preening itself. Indignation and confusion were followed by a certain amount of bluster – and then some loud huffing and puffing preceded another torrent of words: 'Maddie is a strong-minded woman, as you'd expect, but a fine woman – a woman of mature understanding, a woman of refined taste and unqualified candour. There are altercations – of course there are – swords cross, sparks fly. But it's all so much squally weather, the proverbial storm in a teacup. And what am I to do – I ask you – when I am bereft, crest-fallen, outcast – like the dog – on the rooftop – who howls in the night?'

Maddie extended her hand, touched Malcolm's knee, and then slowly withdrew it. The gesture, in spite of its brevity, was tender and affecting.

I saw Malcolm and Maddie for only two more sessions. Maddie returned for a final session – then failed to attend subsequent appointments. I was going to write a letter to the GP anyway, suggesting that they were unsuitable candidates for therapy. To help someone, you have to have a formulation, some idea of what the problem is and how you intend to deal with it. But I had nothing. Moreover, I wasn't even sure that there had ever been an actual problem in the first place. Had Malcolm hit Maddie? I couldn't get him to give me a straight answer no matter how hard I tried. And Maddie was just the same. The GP, on seeing Maddie's bruises, and being disturbed by her oddity, may have overreacted. A consequence of this, perhaps, was that two eccentric (but compliant) people accepted a referral to a psychiatric hospital without question.

Were they personality disordered? They didn't meet diagnostic criteria for any of the specific personality disorders in the DSM. More significantly, neither Malcolm nor

Maddie complained of clinically significant distress. They did argue – and when they argued they both became upset, but not exceptionally so. Clearly, they were odd. And their inability to answer questions was bizarre. But they weren't out of touch with reality; they simply engaged with reality and the social world in a different way from most.

The fascinating thing about Malcolm and Maddie was that they had found each other. Given their unique peculiarities, the likelihood of either of them finding a kindred spirit in the world must have been very low. Yet, somehow – consistent with the romantic ideal – love had found a way. I still wonder how it happened, how they met and what kind of courtship followed. One must suppose that they spent many happy hours ensconced in the Institut Français – before it was overrun by the hoi polloi – discussing gimcrack philosophers and gingerbread catchpennies.

We are all odd behind closed doors. I'm inclined to agree with Alfred Adler, who wisely observed: 'The only normal people are the ones you don't know very well.'

Chapter 12

Brain Cuts: *Love dissected*

The timetable said 'Brain Cuts: Anatomy'. Accompanied by two other students, I went to the dissection room, where a professor with an East European accent welcomed us. By the window a plumber was examining the pipes. The professor lifted the brain from a plastic container and held it under a running tap. He squeezed the formalin out and positioned the gelatinous mass on a marble slab. We sat, like three hungry children, our chairs tilted forward in eagerness. After pointing out the main surface features, the professor made a number of delicate incisions and gently pulled the hemispheres apart. I had heard the sound before in kitchens and restaurants – a succulent tearing. We considered the subcortical structures. Then, after producing a large carving knife, the professor sliced the brain from anterior to posterior, producing a spread of transverse sections. When he bowed his head, we looked at each other furtively and smiled. Each slice of brain revealed interesting patterns of grey and white matter. The cerebellum contained a particularly beautiful branched configuration called the tree of life. The plumber went out to get a screwdriver, but didn't return.

As I sat there, listening to the professor's ripe English, I couldn't help wondering if any memory traces were interred in those slices. Some residual organisation that – with the aid of some impossibly advanced technology – might be translated into moving pictures. I stared at those anatomical slivers with a melancholy fascination, as if my sustained attention might eventually force the dead matter to give up its secrets. I began to imagine scenes from a life, and it was curious, because everything I imagined concerned love and intimacy: a woman lying on rumpled bedsheets – her naked body illuminated by rays of sunshine streaming through a high window; lipstick on the rim of a wine glass; long hair whipped up by a sea-breeze and trained into baroque curlicues against a cloudless sky. Were they still there, in some shape or form, memories like these?

What is life about if it isn't about love? Finding love, being loved and loving others? Yet, love is something we rarely engage with intellectually. We all experience falling in love but take little or no interest in how it works.

When love attracts literary interest, it does so more commonly in genre form. Romantic fiction has never been taken very seriously, and love is diminished even further in the context of romantic comedy, where the standard devices of misunderstanding and cross-purposes make us laugh at the folly of love. More peculiar still is the belief that love is of great importance to women, but of only negligible importance to men. Lord Byron famously said: 'Man's love is of man's life a part; it is a woman's whole existence.' Love is pink with feathered trim, perfumed and mildly diverting, a form of intellectual needlepoint. Frippery.

In reality, love is shaped by Darwinian imperatives. It is another facet of wild nature 'red in tooth and claw'. Falling in love is a combustible state that reproduces the symptoms of psychiatric illness, and when love goes wrong the results can be fatal. Passions can become twisted and ugly.

So why do we joke about love?

Freud believed that we play down those things that make us most anxious. The destabilising nature of love reveals the fragility of the self; in a moment – in the time it takes for eyes to meet across a crowded room – we can lose ourselves. We can become obsessed and mad with desire. A whole life can be overturned and plunge into chaos. And when we consummate love we are humbled. As we explore each other's orifices it is self-evident that we are just animals. We cannot sustain comforting illusions of superiority, cultivation or divinity while exchanging body fluids. The uncomfortable tension created by our contradictory natures – civilised and bestial – is unsettling. No wonder love and its febrile consequences make us anxious.

Love can be other things too – wonderful things. It can be oceanic, transcendent and rapturous. It can make us feel complete. Countless studies show that love – in the context of a satisfying long-term relationship – is associated with well-being and longevity. Love is so powerful it can command Death to come back another day. Conversely, many report that a life without love feels aimless, superficial, lonely and lacking in substance. Love facilitates the tumbling of genes through time, from generation to generation; a never-ending process of recombination that cross-stitches the whole of humanity. It is the greatest of all commonalities.

Love has been selected by evolutionary pressures to maximise reproductive success. Children in the ancestral environment had much more chance of surviving when raised by two committed parents. But how did this situation arise in the first place? What is the *ultimate* explanation for the evolution of love? The answer to this question can be squeezed like a sponge beneath a tap, sliced and spread out in a series of transverse sections.

Compared to most animals, the human infant is appallingly ill-equipped to find food and fend for itself. Somewhat paradoxically, the reason for this extraordinary helplessness is brain size. Intelligence gives human beings an almost incalculable advantage and it has allowed our species to dominate an entire planet; however, enormous brains make childbirth extremely dangerous. Big heads have a tendency to get stuck in the birth canal – babies and mothers die. In evolutionary terms, it is imperative that brain size is optimised, but it is also imperative that offspring survive and reproduce. This problematic contradiction necessitated a compromise. Human babies are born approximately twelve months early compared with other mammals and their brains finish growing outside the womb. Consequently, human babies are extremely vulnerable and must be closely attended. In the ancestral environment, this kind of attention, essential to ensure survival, could be provided only by two parents bonded together for at least three or four years – the time required for offspring to overcome the handicap of prematurity and achieve a level of independence. Parental pair-bonds, although temporary, had to be very strong – overwhelming, in fact – because the future of the human race depended on it. This is why love enslaves us. When we fall in love, we are no longer free. Our capacity for reason is compromised. Our genes don't want us to approach our potential mates with cool detachment. They want us to be on fire – they want us to be impassioned and reckless. They want us to love madly.

If our ancestors hadn't loved madly, their children wouldn't have survived, they would never have reached maturity or made use of their enormous brains, and you wouldn't be reading this book.

Evolutionary pressures selected intelligence, but – in addition to head size – intelligence comes with a further complication. Our big brains allow us to override our instincts. This is a major problem, because what our genes want (reproduction) and what we want (freedom) don't always coincide. A big brain means that we can put our own interests and preferences first. Our ancestors might have chosen in increasing numbers to abandon their partners and children; they might even have chosen to be celibate. This would have resulted in extinction. Evolution compensated for these dangerous possibilities with another compromise. For three to four years at least, the human animal would cease to be an entirely selfish animal. It would be compelled to form an attachment, procreate and care for its offspring.

And all of these objectives were ensured by disabling rational self-interest with what we now call love.

It should be emphasised that evolution is a blind process. There is no master plan and as a result there are many 'unintended' side-effects. Our big brains allow us to take certain liberties with respect to our evolutionary programming. We can, for example, choose to extend pair-bonds well beyond the expedient evolutionary minimum for a multiplicity of very 'human' reasons: companionship, sense of humour, kindness, compatibility, shared memories, an attractive smile, warmth in the middle of the night, eyes of delphinium blue. We all have our reasons.

Over the course of the preceding chapters I have argued for a more considered approach to the subject of love and its relationship to mental health. There is nothing new about this argument, and precedents can be found in numerous classical texts. However, it is my view, largely formed by talking to patients in hospitals and consulting rooms, that we, as a culture, have trivialised an important aspect of the human condition and at a very high cost. We have lost sight of something that Hippocrates, Lucretius and Avicenna incorporated more readily into their world view. A doctor in ancient Greece or Rome, or in eleventh-century Persia, would have had more to say, for example, about lovesickness – in relative theoretical terms – than would a contemporary psychotherapist. Before qualifying as a clinical psychologist, I studied psychology for a total of eight years, during which time I received only one hour's teaching on the subject of romantic love. People can experience significant distress when they fall in love or when love goes wrong. Yet, they are usually wary of talking about their pain openly (especially to a mental-health professional) because they have been made to feel that their predicament is adolescent, foolish or embarrassing – or that their sexual fantasies and urges are dirty or perverted. They are told to get a grip, pull themselves together, or be ashamed of themselves. But it is incredibly difficult to control emotions so deeply rooted in our brains. And, even with professional help, there is no guarantee that the conditions of longing and desire can be successfully treated. More often than not, the goal of therapy is management rather than cure.

One rarely gets an opportunity to look at a brain directly. Nowadays, even neuroscientists have little reason to handle their chosen object of study. With the advent of neuroimaging technology, old-fashioned brain dissections have become increasingly redundant. For me, attending a 'brain cuts' tutorial was curiously haunting. I often think of those slices – gleaming in a pool of autumnal light – and the possibility of residual memories of love lingering, with obscure tenacity, among those florid white and grey patterns.

I wish I had paid more attention to the professor.

The situation and surroundings were oddly allusive with respect to certain cinematic clichés: the hammy East European stresses, the fact that I was sitting in a laboratory, the distinctly Gothic flavour of a human brain being prepared and displayed like a culinary treat. Where did the professor come from exactly? Somewhere quite close to Transylvania, I imagined. Perfect.

Had I been more astute, back then, I might have estimated the professor's age and realised that his accent was meaningful beyond films featuring vampires. I might have determined that a man like him would have an interesting history, the kind of history that would have deepened and enriched my philosophical reflections on the brain, life and love. I might have delayed my departure after the tutorial, tarried, asked more questions and engaged him in a broader conversation. But I didn't. Regrettably, my curiosity took me only as far as Transylvania and the looming turrets of Castle Dracula.

Recently, I discovered that the anatomy professor was a Holocaust survivor. He had spent a winter in Bergen-Belsen concentration camp and his father had died there. The professor had been a child at the time. I shudder – even now – to think of what memories were preserved in his grey matter.

Life is a precarious business and love is its essential ingredient. As I get older, I have found it salutary to remind myself of these truisms with increasing regularity. We humans – it seems to me – are inclined to forget the obvious.

'I hate him. I hate him.'

Verity was a middle-aged stockbroker's wife from a wealthy, upper-crust family, a former debutante whose life was a continual round of coffee mornings, village fetes, gymkhanas, Wimbledon, charity events and trips to see the opera at Glyndebourne. Her children had grown up and left home but that hadn't left her feeling bereft of purpose. Life was good. In fact, life had always been good.

Without any prior indications and completely unexpectedly, her husband had announced that he was no longer happy. He had decided to move to his own house in rural

Kent and wanted a divorce. When Verity asked if there was another woman, her husband had replied that there wasn't. Verity didn't believe him. Conventional methods of detection failed to resolve the issue, so Verity opted for more drastic measures. She swapped her expensive gowns and floral dresses for combat fatigues and spent nights camped out in a field near her husband's new home. She purchased high-powered binoculars, a camera with a telephoto lens, and a long-distance listening device. Verity was no longer acting like a sophisticated society hostess and mother of four. She was acting like a secret service operative on a mission behind enemy lines. It had been an extraordinary metamorphosis. Her friends thought she had gone mad.

Verity was sitting cross-legged on her hospital bed. I was sitting in front of her on a wicker chair. A biography of Margaret Thatcher and an Agatha Christie novel were positioned on the bedside cabinet next to a jug of dusty water and a paper cup. She was wearing a baggy cardigan and track-suit bottoms. Her hair was mussed and her face was deeply lined. She had lost a lot of weight in a short period of time and this had produced a pendant wattle beneath her chin.

'I hate him,' she repeated bitterly.

As it turned out, there *was* another woman – and that woman was young and beautiful. Verity hadn't been able to stop watching. She had wanted to know everything about her husband's new life, and very soon she did. It was then that her sense of empowerment crumbled and she became clinically depressed.

'I just wanted to know the truth. That's what I told myself.

I didn't want him to get away with lying to me – I felt he was insulting my intelligence. I'm not sure why I carried on. The whole thing became compulsive, morbid.' She clutched her head, as if her brain had been sliced into sections and she was trying to stop it from falling apart. 'I don't know how it came to this.' She glanced around the room, acknowledging with frightened eyes that she was in a psychiatric hospital. 'I don't know who I am any more.' She cried for a while and I offered her some tissues.

After Verity had dried her eyes she said, 'The girl – I think of her as a girl – she was oriental – Chinese – and . . .' Misery withered the end of her sentence. 'People think I'm ridiculous. It'll get around, of course, these things always do. People can be very insensitive and indiscreet. But am I *so* ridiculous?' She cupped her chin, thoughtfully. 'I say I hate him. But I know that's not right.' Cinders of memory were rekindled by a heavy sigh: a hotel bedroom – a restaurant in Paris – a beach walk on a blustery day, perhaps? 'If I really hated him it wouldn't hurt this much. It only hurts this much because . . .' The next words would be difficult and the strain showed on her face. 'Because I still love him.'

That was what I was waiting for: self-recognition, truth – something I could work with. Now we could begin.

Acknowledgements

Thanks to: Richard Beswick for encouraging me to write this book and being a creative, astute and enthusiastic editor; my agent Clare Alexander; Nicola Fox (for reading earlier drafts); and Nithya Rae for a thoughtful and impressive copy-edit.

Mental health organisations, societies and charities in the UK

The Incurable Romantic includes descriptions of several patients who were difficult to treat. This was done intentionally – psychotherapy can be challenging, and it is impossible to guarantee a cure. That said, when psychological treatments work they can be transformative. In fact, even when they don't work patients can still benefit enormously from management advice and ongoing support. Access to psychological therapies is usually achieved by referral through a GP. If you are affected by any of the psychological problems mentioned in this book, the following contact list might be useful.

General mental health

Samaritans
www.samaritans.org

Mind
www.mind.org.uk

Sane
www.sane.org.uk

Addiction

Addaction
www.addaction.org.uk

Association for the Treatment of Sex Addiction and Compulsivity
www.atsac.co.uk

Eating disorders

Anorexia and Bulimia Care
www.anorexiabulimiacare.org.uk

Anxiety

Anxiety UK
www.anxietyuk.org.uk

Obsessive-compulsive disorder

OCD Action
www.ocdaction.org.uk

HIV

Terrence Higgins Trust
www.tht.org.uk

Relationships and domestic violence

Relate
www.relate.org.uk

National Domestic Violence Helpline
www.nationaldomesticviolencehelpline.org.uk

Sexual abuse

The National Society for the Prevention of Cruelty to Children
www.nspcc.org.uk

Ageing

Age UK
www.ageuk.org.uk

Finding a qualified therapist

British Psychological Society
www.bps.org.uk

ABACUS